Money and Finance After the Crisis

Antipode Book Series

Series Editors: Vinay Gidwani, University of Minnesota, USA and Sharad Chari, University of California, Berkeley, USA.

Like its parent journal, the Antipode Book Series reflects distinctive new developments in radical geography. It publishes books in a variety of formats – from reference books to works of broad explication to titles that develop and extend the scholarly research base – but the commitment is always the same: to contribute to the praxis of a new and more just society.

Published

Money and Finance After the Crisis

Critical Thinking for Uncertain Times

Edited by

Brett Christophers, Andrew Leyshon
and Geoff Mann

WILEY Blackwell

Registered Office(s)
John Wiley & Sons, Inc., 111 River Street, Hoboken, NJ 07030, USA
John Wiley & Sons Ltd, The Atrium, Southern Gate, Chichester, West Sussex, PO19 8SQ, UK

Editorial Office
9600 Garsington Road, Oxford, OX4 2DQ, UK

For details of our global editorial offices, customer services, and more information about Wiley products visit us at www.wiley.com.

Wiley also publishes its books in a variety of electronic formats and by print-on-demand. Some content that appears in standard print versions of this book may not be available in other formats.

Library of Congress Cataloging-in-Publication Data

Names: Christophers, Brett, 1971– editor. | Leyshon, Andrew, editor. | Mann, Geoff, editor.
Title: Money and finance after the crisis : critical thinking for uncertain times /
   [edited by] Brett Christophers, Andrew Leyshon, Geoff Mann.
Other titles: Antipode book series.
Description: First edition. | Hoboken, NJ : John Wiley & Sons, 2017. |
   Series: Antipode book series | Includes index.
Identifiers: LCCN 2017009013 (print) | LCCN 2017013973 (ebook) |
   ISBN 9781119051428 (cloth) | ISBN 9781119051435 (pbk.) |
   ISBN 9781119051404 (pdf) | ISBN 9781119051398 (epub)
Subjects: LCSH: Global Financial Crisis, 2008-2009. | Banks and banking. |
   Economic history–21st century.
Classification: LCC HB3717 2008 .M674 2017 (ebook) | LCC HB3717 2008 (print) |
   DDC 332–dc23
LC record available at https://lccn.loc.gov/2017013973

Cover Image: © leszekglasner/Gettyimages
Cover Design: Wiley

Set in 10.5/12.5pt Sabon by SPi Global, Pondicherry, India

Contents

Series Editors' Preface

The *Antipode Book Series* explores radical geography 'antipodally,' in opposition, from various margins, limits or borderlands.

Antipode books provide insight 'from elsewhere', across boundaries rarely transgressed, with internationalist ambition and located insight; they diagnose grounded critique emerging from particular contradictory social relations in order to sharpen the stakes and broaden public awareness. An *Antipode* book might revise scholarly debates by pushing at disciplinary boundaries, or by showing what happens to a problem as it moves or changes. It might investigate entanglements of power and struggle in particular sites, but with lessons that travel with surprising echoes elsewhere.

Antipode books will be theoretically bold and empirically rich, written in lively, accessible prose that does not sacrifice clarity at the altar of sophistication. We seek books from within and beyond the discipline of geography that deploy geographical critique in order to understand and transform our fractured world.

Vinay Gidwani
University of Minnesota, USA

Sharad Chari
University of California, Berkeley, USA

Antipode Book Series Editors

Notes on Contributors

Dick Bryan is Emeritus Professor of Political Economy at the University of Sydney. His current research concerns the reframing of 'safe' assets in an era of 'unsafe' treasury bonds and with the ways in which imminent developments in blockchain technology might impact on value theory.

Brett Christophers is Professor of Human Geography at Uppsala University. His most recent books include *Banking Across Boundaries: Placing Finance in Capitalism* (2013; also in the Antipode Book Series) and *The Great Leveler: Capitalism and Competition in the Court of Law* (Harvard, 2016). His textbook, *Economic Geography: A Critical Introduction*, co-authored with Trevor Barnes, will be published in 2018 by Wiley-Blackwell.

Joseph A. Daniels is a Joint-PhD student in the Department of Geography at the University of British Columbia and the School of Geography at The University of Nottingham. His current research unpacks the cultural political economy of crowdfunding, with wider research interests in 'alternative' economic practices, financial geography, and urban political economy. His past work has focused on the role of bank restructuring and the financialization of real estate in transforming Singapore's space-economy.

Jessica Dempsey is an Assistant Professor in the Department of Geography at the University of British Columbia, Canada. Her research interests include biodiversity politics, ecosystem services, and green finance, drawing from diverse methodological approaches and literatures including economic geography, feminist political economy/science studies, and political ecology. Her first book *Enterprising Nature* is also on the Antipode book series (2016).

Gary A. Dymski received his PhD in economics from the University of Massachusetts, Amherst in 1987. He joined the faculty at the Leeds University Business School (LUBS) as Chair in Applied Economics in 2012 after twenty-one years in the University of California system. Gary's research focuses on discrimination and redlining in credit markets, urban economic development, financial crisis, the subprime and Eurozone crises, banking and financial regulation, and urban development.

Marieke de Goede is Professor of Politics at the University of Amsterdam. She has written widely on preemptive counterterrorism and the role of financial data. She is author of *Speculative Security: Pursuing Terrorist Monies* (2012) and co-editor of the special issue on 'The Politics of the List' of *Environment and Planning D: Society and Space* (2016). De Goede is principal investigator of FOLLOW: Following the Money from Transaction to Trial, funded by a European Research Council grant. She is Associate Editor of *Security Dialogue*.

Deborah James is Professor of Anthropology at LSE. She is author of *Money from Nothing: Indebtedness and Aspiration in South Africa* (Stanford University Press, 2015), which documents the precarious nature of both the aspirations to upward mobility and the economic relations of debt which sustain the newly upwardly mobile in that country.

Karen P.Y. Lai is Assistant Professor of Geography at the National University of Singapore. Her research interests include geographies of money and finance, markets, service sectors, global city networks and international financial centres. Her current project examines the global financial networks of investment banks in mergers and acquisitions, and initial public offerings. She is on the Standing Committee of the Global Production Networks Centre at NUS, and editorial board member of *Geography Compass*.

Paul Langley is Professor of Economic Geography at Durham University, UK. His research to date has developed through the publication of three monographs: *World Financial Orders* (Routledge, 2002/2013), *The Everyday Life of Global Finance* (Oxford University Press, 2008), and *Liquidity Lost* (Oxford University Press, 2015). His current research addresses novel forms of digital and/or social finance that have consolidated in the wake of the global financial crisis, such as impact investment, crowdfunding and peer-to-peer lending.

Andrew Leyshon is Professor of Economic Geography and Associate Pro-Vice Chancellor for Research & Knowledge Exchange in the Faculty of Social Sciences at the University of Nottingham. His work has mainly focused on money and finance, the musical economy and the emergence of diverse economies. Books that reflect these interests include *Money/ Space* (with Nigel Thrift, Routledge, 1997), *Reformatted: Code, Networks and the Transformation of the Music Industry* (Oxford University Press, 2014), and *Alternative Economic Spaces* (with Roger Lee and Colin Williams, Sage, 2003). He is a member of the Editorial Board of *Environment and Planning* and *Journal of Cultural Economy*, and is on the Editorial Advisory Board of *Economy and Society.*

Geoff Mann is Professor of Geography at Simon Fraser University in British Columbia, Canada. His research focuses on the theory and politics of economic governance in liberal capitalism, especially as it concerns income distribution, poverty and unemployment. His most recent book is *In the Long Run We Are All Dead: Keynesianism, Political Economy and Revolution* (Verso, 2017). He is also the author of *Disassembly Required: A Field Guide to Actually Existing Capitalism* (AK Press, 2013), and *Our Daily Bread: Wages, Workers and the Political Economy of the American West* (University of North Carolina Press, 2007).

Phillip O'Neill is the Director of the Centre for Western Sydney at Western Sydney University. He is an economic geographer with expertise in infrastructure financing. Besides academic publications, Phillip is a leading contributor to policy for the economic management of cities in Australia. He is also a prominent public commentator and newspaper columnist.

Michael Rafferty teaches within the International Business Programme at the College of Business, RMIT University. His research engages with the changing organizational forms of global capital and the increasingly financialized logic that informs and commensurates those processes.

Duncan Wigan is Associate Professor In International Political Economy at the Department of Business and Politics, Copenhagen Business School. His research focuses on issues of international taxation and international finance. In 2018, with co-author Leonard Seabrooke, he will publish an edited volume, *Global Wealth Chains: Managing Assets in the World Economy* (Oxford University Press) and, also with Leonard Seabrooke, a monograph, *Global Tax Battles: The Fight to Govern Corporate and Elite Wealth* (Oxford University Press).

1

Money and Finance After the Crisis

Taking Critical Stock

Brett Christophers, Andrew Leyshon and Geoff Mann

Introduction

This is not another book about the financial crisis, or not exactly. It is, rather, a multidisciplinary collection of essays that dwell on the geographies of money and finance unfolding in its wake – the dynamic and sometimes volatile post-crisis financial and monetary worlds we inhabit. Indeed, the accounts of these geographies are perhaps best thought of as 'worldings': like the crisis itself, the contributions touch upon an extraordinary range of contemporary lifeworlds and social formations, including not only the complexities of modern debt-driven financial markets, but also their fascinating and unexpected connections to (for example) post-Apartheid South Africa, the 'War on Terror' and biodiversity conservation. They reveal, if we did not already know it, that there is neither a universal experience of the current situation, nor a single 'correct' analysis, nor is there any 'representative' agent, place or scale that captures its dynamics. The breadth and complexity of our monetary-financial conjuncture demands that we assemble a collection of a breadth and specificity adequate to it.

In addition to the benefits of the range of insights these chapters offer into the processes of financialization and the practices and imaginaries through which it proceeds, bringing them together in one collection illuminates dynamics and questions that would be much less visible without their juxtaposition. One of the most important of these, perhaps,

Money and Finance After the Crisis: Critical Thinking for Uncertain Times,
First Edition. Edited by Brett Christophers, Andrew Leyshon and Geoff Mann.
© 2017 John Wiley & Sons Ltd. Published 2017 by John Wiley & Sons Ltd.

concerns the status of the concept of 'crisis' itself. For it seems fair to say that in the serial nature of 'the' crisis – which in the simplest sense began in 2007–2008 in a specific set of asset markets in the United States, but has since continually rolled over into other markets, places, and realms of financialized modernity – the term 'crisis' has lost some purchase. This is not a merely 'academic' problem, and not only because such conditions make it very difficult to determine how, exactly, the financial crisis was made manifest: was it a crisis of the financial sector? Or of the state? Or capitalism? Perhaps neoliberalism? What about money or monetary governance? Or (almost certainly) some combination thereof? We must also consider the fact that if seemingly everything monetary or financial (and much else besides) appears to be always 'in crisis', or to sit permanently on the edge of a precipice, how much analytical help is the concept of 'crisis' anymore? In an historical moment we might be forgiven for confusing with Walter Benjamin's permanent state of emergency, it is the concept of 'crisis' that is in crisis. The 'beginnings' and 'ends' of crises are relative, not absolute; the solution of a crisis for some often means it has been passed on in a repackaged and repurposed form for others to bear (Christophers 2015a).

If so, the problem we confront is not simply terminological, but existential. The challenge is not just finding another category that can help us rearrange the conceptual bits and pieces. Even if we pretend that the time before the collapse of subprime mortgage-backed securities markets and the bankruptcy of the New York investment bank Lehman Brothers – two of the most commonly-identified 'beginnings' of the crisis – was somehow an untroubled era of normality (which would of course be absurd), we are forced to acknowledge that the monetary and financial geographies produced since then are not accurately described as 'post-crisis' geographies, as if 'the' crisis happened, ended, and now we live in its 'aftermath'. We do not inhabit 'post-crisis geographies', but 'crisis geographies' – spaces, places, imaginaries and practices – that have been and continue to be constituted in and by crisis.

Which means not only is it no longer necessarily possible to analytically isolate or recognize 'crises' by the specificity of their dynamics – which would rely on an old model in which crisis is by definition identifiable as an exception to 'normality' – but also that one of the key categories through which we narrate modernity has lost much of its concrete reference (Kosselleck 2006). In other words, if 'crisis' is both foreground and background, how do we identify it when we see it? If capitalism is defined by its tendency to crises, what does it mean when crises are no longer just tendential, but endemic, even constitutive?

The challenges these questions place before us cannot be dismissed, and as such they merge the analytical problems with the existential.

Money and finance are no longer 'containable' to their proper spheres, if they ever were. They are, today, some of the principal means by which 'crisis' becomes remarkably unremarkable, normal even, because immanent to both money and finance is a contradictory sense of possibility, one that always carries a potential 'solution' and a 'dissolution' to the future. In Kurt Vonnegut's *God Bless You, Mr. Rosewater* (1965, p. 171), a character defines money as 'dehydrated utopia', which in some respects it certainly is. But it is also, no less certainly, dehydrated disorder; a little bit of rain and it is apparently as likely to precipitate calamity as it is utopia – and it does so in geographically and historically specific ways. What can scholarship contribute at such a conjuncture? In some ways, it is as simple as keeping one's feet on the ground – in the world – at a moment when the temptation to exercise analytical control leads so many to unmoored abstraction. As the poet Anne Carson (2016, p. 4) put it, with precisely these challenges in mind, 'What can you control? Wrong question. Can you treat everything as an emergency without losing the reality of time, which continues to drip, laughtear by laughtear?'

The chapters in this book attempt something like this: to approach the (post-) crisis world as a collection of emergencies without losing the reality of time – *and* space. They illuminate the complex and often unpredictable variations on a monetary and financial theme, from the realms of orthodox finance capital to biodiversity conservation. The chapters were all written several years after the moment of 2007–2008, and draw attention to the significance of thinking about money and finance geographically, especially given that the roots of the crisis took hold in distinctive and interlinked geographies. These ranged from the global financial centres of New York and London, wherein financial innovation and alchemy was applied to ever more complex and interlayered financial products, through the surplus economies of Asia and the Middle East that provided the financial boom that preceded the crash with so much momentum and financial weight, to the everyday markets that shape, to no small extent, both the financial realm of investment and debt opportunities, and the vagaries of our collective and individual fortunes.

The remainder of this introductory chapter is organized as follows. In the section 'The Crisis and the Academy: "I told you so"' we explore the relationship between academic work and the financial crisis, focusing in particular on the constitutive role played by the discipline of economics in justifying the reregulation of markets and institutions that preceded the crisis, but also on the response to the crisis in that discipline and in cognate social sciences. Next, in 'The Crisis, the State and Regulation: "What have I done?"' we focus on the role of states, policy makers and regulators in the period preceding and following the crisis. 'The Crisis

and the Financial Sector: "What shall we do now?"' looks at the implications of the crisis for the financial services sector. The final section,
'The Content of the Book', provides a summary of the chapters that
make up the rest of this volume.

The Crisis and the Academy: 'I told you so'

The political-economic upheaval that began in 2007–2008 has, in the
years since, fundamentally reshaped research trajectories across the
social sciences. For mainstream economics the implications are, of
course, potentially soul-shattering, what Mirowski (2010; 2013) labels
the 'Great Mortification'. Even those who still refuse to abandon the
fortress of orthodoxy that defended the forces of financial chaos enjoy
little peace, compelled as they have been to pull up the drawbridge and
self-righteously repel wave after wave of critique from fellow-econo-
mists and non-economists alike (e.g., Lucas 2009). But the new 'economic
reality' has effected significant, if less existential, changes in other social
sciences too (as well as in the humanities, though that is not our focus
here). In this, the extraordinary scholarly breadth of its impacts, the first
major capitalist crisis of the twenty-first century is quite different than
the Great Depression of the 1930s or the collapse of the post-World War
II 'Long Boom' in the 1970s.

Of course, both those moments were important objects of scholarly
analysis, and contemporary interest certainly extended beyond the halls
of economics departments. Poverty, unemployment, homelessness, the
relation between the state and markets – all of this was, unsurprisingly,
on everyone's radar. But – to borrow a phrase from one of the most
celebrated of orthodox economics' recent apologias (Reinhart and
Rogoff 2009) – on the wide scholarly front, this time *is* different, in one
crucial sense: never before have so many non-economists delved so
deeply into the 'technical' details of modern (capitalist) economic science, policy and practice. Social scientists of all stripes have found the
formerly impenetrable and even uninteresting fortress of economics and
finance irresistible and, if not captured it, have certainly gathered a lot of
intelligence about the goings on inside the walls. Geographers, anthro-
pologists, sociologists, psychologists, historians, political scientists – all
have in the last less-than-a-decade, written, often wisely and compel-
lingly, on such once-esoteric subjects as mortgage securitization
(Walks and Clifford 2015), credit risk (Ashton 2011), and investment
strategy (Hansen 2015). Topics that might once have anaesthetized most
non-economists, like the yield curve and accounting practices, have
(justifiably) become hot topics (Christophers 2017; Joseph 2014;

Zaloom 2009). Economics and 'finance' in particular, are now, more than ever, more than economic.

It must be said that this explosion in scholarship – especially in its so-called 'critical' variety – is partly caught up in a somewhat irrepressible *Schadenfreude*. Almost every scholar outside orthodox 'neoclassical' economics has secretly (or not so secretly) enjoyed the crisis in the self-described 'queen of the social sciences' (Samuelson 1973, p. 6). Many, heterodox economists in particular, have understandably been unable to suppress the urge to say 'I told you so' (Wolff 2010; Taylor 2010; Mirowski 2013). But that is hardly the entirety of the non-mainstream contribution. Scholars across a variety of social science disciplines have discovered they actually have much to say about finance, and with economics' common sense seemingly proven so wrong by the continued unravelling of the global economy via financial meltdown, now they have an audience. This multidisciplinary project of unpacking money and finance has generated fascinating and crucial insight. While not suggesting that there are no pre-crisis foundations upon which to begin, one might even say it has shifted how we understand modernity, locating money and finance – social relations like valuation, debt-credit, and securitization – much nearer the centre of our conceptual consciousness.

If, for convenience's sake, we mark the full-blown implosion of the global financial system with Lehman Brothers' bankruptcy on 15 September 2008, we can retrospectively identify a few different developments in post-Lehman scholarship on money and finance. Of course, no such account can be comprehensive, if for no other reason than the dynamic and diverse forms 'crisis' is deemed to have taken in the intervening years, from 'subprime' to 'credit crunch' to 'sovereign debt' to 'Eurozone' to 'emerging markets', among others. Any path we try to map will always leave something out, suggesting a coherence within what is in fact a broad collection of many more-or-less related efforts we can only trace at a general scale. One of these scholarly dynamics, of course, is mostly internal to orthodox economics, in which 'progressive' forces batter away at a still-hegemonic old guard, while a 'reasonable' middle ground tries to mediate. A second is in what the American Economic Association (AEA) calls the 'allied social sciences', i.e., largely heterodox and radical economics and political economy, economic geography, economic sociology and political science. We can see a third dynamic in the more emphatically 'humanistic' cultural economy research in human geography broadly defined, anthropology, sociology, cultural studies and political and social theory. Again, however, all these developments impact each other, so telling a story that is too organized or clear would be a misrepresentation.

Perhaps the only thing we can know for certain is that the debates in economics have much more in common with those in 'allied' fields of knowledge, while these 'allies' tend to be in closer conversation with fields outside economics proper than orthodox economics itself, which generally isolates itself as much as possible, and dismisses critique from outside. This has only exacerbated one of the more fascinating scholarly effects of the crisis. For if Lehman's disintegration signals the bursting of the asset bubble that had inflated over the preceding quarter century at the heart of the richest economies on the planet, 2008 also marks the shattering of economics' disciplinary ego at its most distended. The burgeoning hubris of orthodox macroeconomics during the so-called Great Moderation of the 1990s and first half of the 2000s, hardly bothered by the Asian financial crisis of the late 1990s and the dot.com crash of the early 2000s (the rapid 'resolution' of which merely increased the mainstream's confidence), led to historically unprecedented self-congratulation (Fourcade, Ollion and Algan 2015). Sharpened by the rational expectations 'revolution' and a 'new neoclassical synthesis' captured in dynamic-stochastic general equilibrium models, macroeconomics declared itself virtually complete. Nobel-winner Robert Lucas, a key contributor to these technical developments, announced in his 2003 Presidential Address to the AEA that 'the central problem of depression prevention has been solved, for all practical purposes, and has in fact been solved for many decades' (Lucas 2003, p. 1). Four years later, itemizing mainstream economics' 'tremendous advances in knowledge' on the very eve of calamity, Harvard's Benjamin Friedman (2007, pp. 49–50) spoke to his colleagues of a 'well-earned complacency', since the 'past quarter-century has been about as good a run, at least in aggregate dimensions, as one is likely to get'.

Let us just say these declarations proved premature, and turn our attention to the unavoidably existential disciplinary implications that might follow a fall from such confident heights. Although it is simply not possible to do justice to the range of debates that have occupied economics since 2008, even if we narrow our focus to the subfields of macroeconomics, monetary or financial economics, one notable – and crucial – feature is the way in which the debate among economists has been almost entirely confined to the technical realm. The problem – whether it be leverage, risk, expectations, liquidity, employment or anything else you can think of – is always with the models or with policy (see Gorton and Metrick 2012; Lo 2012). Critique has thus focused on 'scientific' questions, like the so-called 'Gaussian copula' in the risk models of subprime asset-backed securities, which underplayed the likelihood of mass mortgage default, or on structural obstacles to policy shifts like 'labour market reform' (i.e., lower wages and reduced social protection). The fix in either case is technical: better models, different policy, institutional tweaks (for example, Swagel 2015; Thimann 2015). This is true not only of the analyses of those fully committed to the status quo, but also of some of the most well-known internal 'critics' of economics' methods and ways of knowing (Rodrik 2015). Which is to say, as Rick Wolff (2009, p. 3) put it, for the vast majority of economists, the current problem *in* capitalism is not a problem *of* capitalism.

This is in no way to suggest technical questions are trivial or pedantic. For example, another, arguably much weightier, 'technical' question arose in the furore around econometric estimations of sustainable levels of sovereign debt (note that the very possibility of a furore over econometric modelling is arguably a function of the crisis: the controversy reached the op-ed pages of the *New York Times* and the *Financial Times* and was closely covered by the *New Yorker*). In a series of contributions beginning in 2009, influential orthodox economists Carmen Reinhart and Kenneth Rogoff turned their attention to public debt levels in an effort to understand the precarious situation of many states' finances following the meltdown and the ensuing banking crises. Their 2010 paper 'Growth in a Time of Debt' (Reinhart and Rogoff 2010) concluded that countries with public debt levels exceeding 90 per cent of annual gross domestic product (GDP) were prone to substantially lower growth than countries carrying debt below the 90 per cent threshold.

While Reinhart and Rogoff's research is of course not solely responsible for the imposition of austerity upon many polities since the beginning of the crisis, their energetic promotion of the paper and its austere implications definitely helped it along. In the Greek and Spanish debt crises, for example, it was a key piece of evidence for those forces insisting on strangling public finances, imposing crushing levels of taxation on the working and middle classes, and reducing or rescinding promised social support. That the stakes in such claims are not small is proof that 'technical' concerns are not necessarily minor. Indeed, it was no minor matter when Thomas Herndon and colleagues performed a 'critical replication' of the Reinhart–Rogoff analysis in 2014, uncovering errors and selective data exclusion in the original which completely biased the results, and showed there was no such thing as the 90 per cent debt-to-GDP threshold (Herndon, Ash and Pollin 2014). If the Troïka (the European Commission, the European Central Bank and the International Monetary Fund) had acknowledged that Reinhart and Rogoff's model, crucial to their austerity programme (Kumar and Woo 2010; Gongloff 2013), was founded in the same common sense that endorsed the dynamics that led to the financial collapse, or at least been willing to consider this alternative economic analysis, things would arguably look very different in southern Europe today (Chick and Dow 2012; Stiglitz 2013b).

The most common alternative economic analysis in the post-Lehman era, on which Herndon et al. and many other economists have been working, is best described as loosely Keynesian. The resurrection of Keynes and the explosion of Keynesian enthusiasm almost immediately upon the onset of financial chaos in 2007–2008 was firmly grounded in a scathing critique of mainstream economics (Backhouse and Bateman 2011; Eatwell and Milgate 2011; Skidelsky 2009; Taylor 2010; Temin and Vines 2014). While the uses of the term 'Keynesian' in the discipline of economics are quite varied, and often at odds with the 'welfare state' associations shared by most non-economists, at its broadest it describes analyses based on two fundamental axioms concerning 'actually exist-ing' modern economies: first, there is no reason to expect economic activity to lead 'naturally' to full employment; and second, insuperable uncertainty about the future always shapes the present (Mann 2017; cf. Dymski, this volume). Both of these conditions are dismissed a priori by the orthodox economics to which Reinhart and Rogoff cling. That orthodoxy begins from the assumption that full employment equilibrium is 'free' markets' inevitable *telos*, and works on the basis of 'rational expectations', which is to say that economic actors are modelled as if they enjoyed what one might call adaptive omniscience. According to the theory of rational expectations, nothing is unexpected; even surprise is unsurprising: when Queen Elizabeth II (among others) asked during a visit to the London School of Economics why economists had failed to anticipate the economic wreckage, Lucas responded (in a comment in the *Economist*) that economics predicts that we cannot predict such events (Kay 2011).

The Keynesian revivalists are a relatively diverse bunch. They range from social democratic 'post-Keynesians' like Herndon, Ash and Pollin (2014) through pretty-close-to-orthodox 'New Keynesian' thought-leaders like Paul Krugman (2012) and Joseph Stiglitz (2010), to former champions of neoliberal deregulation like Richard Posner (2009) and the *Financial Times*' Martin Wolf (Wolf 2008a; Wolf 2008b). All gener-ally agree on the most important lesson economists should take from the crisis: that coordinated public authorities need to keep a much closer eye and hand on the level of risk and volatility in the economy (and be ready to jump in to stabilize quickly and comprehensively) because, contrary to the assumptions of the neoclassical rational expectations 'revolution-aries' and their models, the future is not well-characterized by a series of calculable risk probabilities, but rather by the fundamental, qualitative uncertainty that Frank Knight (1921) famously distinguished from risk. When banks stopped lending in the 'credit crunch', or when bond mar-kets would not touch Greek debt during the Eurozone's 'sovereign debt crisis', it was because the only thing that might give them confidence – state and multilateral commitment to a stable future – was not forthcoming (DeLong and Summers 2012).

Even though Keynes' name was only briefly a commonplace – overwhelmed, some say, by the restoration of orthodox neoclassicism by 2011 – these broadly Keynesian ideas in fact had an enormous impact, if not in the programmatic and sustained manner some Keynesians called for (Stiglitz 2013a). For instance, the increasingly common phenomenon of negative interest rates (Sandbu 2016) and quantitative easing across Europe and the US (Palley 2011) are both premised on Keynesian thinking (even though they represent alternative approaches to the same problem, i.e., lack of investment demand). Even if the fiscal support system that many (unlike Keynes himself) associate with Keynesianism has not been reconstructed, macroeconomics and macro policy have nonetheless been deeply informed by Keynesian economists' diagnosis of the crisis. Whether this diagnosis enables states and elites to avoid similar collapses in the future or not, it is clear that many economists have come to embrace the Keynesian wisdom that we have very little idea what lies ahead, and things will not look after themselves, or certainly not soon enough.

This broadly 'Keynesian' realm is really an overlapping transition zone between the realm of mainstream economics and that of heterodox and radical economics and political economy, and economic geography and sociology. Indeed, outside the work of dissenting economists who, in the tradition of Hyman Minsky or Nicolas Kaldor, rely on mainstream methods and concepts but turn them to oppositional effect (for example, Seccareccia and Lavoie 2015; Taylor 2010), it is often difficult to identify any specifically disciplinary differences in this group of scholarly analyses of the crisis: economic geography can be indistinguishable from radical political economy (Harvey 2014) for example, or economic sociology (Krippner 2011) can frequently pass for post-Keynesian political economy (Schmidt 2009). What differences persist regarding the crisis lie less in the explanation of what went wrong than in the account of longer-term political-economic and social transformation in which the crisis is situated.

The range of such differences is usefully captured in the diverse reactions to what is by far the best-known scholarly product of the crisis, Thomas Piketty's *Capital in the Twenty-First Century* (Piketty 2014). While Piketty's argument is not focused specifically on the crisis-as-event, but rather on capitalism's instability because of inherently disequalizing forces (as described by Piketty's now famous inequality, $r > g$, i.e., return on capital generally exceeds the growth rate, so the 'haves' surge ahead as the 'have-nots' fall further and further behind), *Capital in the Twenty-First Century* is in many ways economics' crisis-book par excellence for the crisis world. It was motivated by, and appeared in, a post-Lehman world obsessed with the causes and consequences of dynamics the crisis seemed to illuminate so clearly – financialization and seemingly absurd levels of finance-driven inequality, stagnating incomes and credit-based consumption, and the retreat of the state into deregulatory irresponsibility. However well written, it is safe to say that a massive tome sprinkled with graphs of economic growth rates would not have been a best-seller without the financial crisis. Its implicit moral economy, focused on the unequal (or even unearned) distribution of rewards, resonated with popular debates concerning the responsibility for the crisis and its social costs.

An eminent mainstream (if politically progressive) economist, Piketty understood *Capital in the Twenty-First Century* as a contribution to the debate in his discipline. And certainly it is; the book attracted extraordinary attention from economists, and in the process lit up the divide between Keynesians (of all stripes), who mostly celebrated it (Krugman 2014; Summers 2014), and the orthodox faithful, who denounced its conclusions, methods, data and so on (e.g., Mankiw 2015, or the dismissive reviews in the ultra-free-marketeering *Journal of Political Economy* (Blume and Durlauf 2015; Krusell and Smith 2015)). But the non-mathematical framing of the book, its historical sweep, and the macrosocial scale of its focus meant that it was also embraced enthusiastically if not uncritically among non-economists inside and outside the academy. In fact, it exemplifies a burgeoning trend in non-economists' analyses of the crises that have played out since 2007–2008: a turn from a fascination with the 'triggers' – the tools and techniques of modern finance like special purpose vehicles, or mark-to-model valuations – to an increasingly longer-run, structural or even 'epochal' understanding of the crisis (Postone 2012; Adjoran 2014). It is almost as if the feeling of wandering the corridors of the palace after the revolution, marvelling at the deceit and luxury now visible for all to see – epitomized by the films *Inside Job* and *The Big Short* and so common among critics of finance after the collapse of Lehman – has diminished as time has passed, and the *longue durée* of capitalist and (neo)liberal social relations is attracting widespread attention again (Fraser 2013; Panitch and Gindin 2013; Moore 2015; Christophers 2016b). Indeed, as discussed in the opening paragraphs, we should not be surprised that some are asking what exactly is in crisis, and how 'crisis' became so central to our conception of history (Roitman 2014; Fraser 2015).

This development has had the interesting effect of linking the general critical trend to the radical and Marxian literature to an unprecedented degree. As one might expect, Marxists were distinctive in the early moments of the crisis for their welcome effort to look past the machinations of financial wizardry to understand the subprime crisis and credit crunch as the latest form of capitalism's ineradicable crisis tendency (Foster and Magdoff 2009; Duménil and Lévy 2011; McNally 2001). They pointed to the contradictions built into credit-based neoliberal political economies and the squeeze on workers in the wake of their relative successes during the post-World War II 'Long Boom', and positioned the present moment as only one more instantiation of a process at work for two centuries. In its broad outlines, this analysis has gained a great deal of traction across the social sciences; 'critical' scholars who would never have considered themselves Marxists have nonetheless embraced the accounts of committed Marxists like David Harvey (2010) and Leo Panitch and Sam Gindin (2013).

Yet, as compelling as many of these analyses are, it must be acknowledged they are not enough on their own. Since capitalism is crisis-producing, even increasingly so – an argument that can no longer be attributed solely to Marxists and Keynesians (Mann 2015), but rather now seems simple common sense – these analyses are of course essential. At the same time we must recognize that the specificity of crisis as a geographical and historical process – what is qualitatively and quantitatively particular to these times and these many places – can get lost or forgotten if its 'empirics' are reduced to merely the latest variation on a universal theme, whether it be accumulation and exploitative overaccumulation or $r > g$.

Even if such accounts of inevitable structural contradiction are perfectly accurate, they often tend to diminish or leave underexamined no small part of what matters: the widely varied and often harrowing experiences and implications of this particular complex of crises for the many different and overlapping groups that make up contemporary societies. Moreover, while efforts to concretize crisis 'in the world' by emphasizing histories of class struggle or examining specific national or international 'contexts' – of which there exists a long and laudable tradition (e.g., O'Connor 1973; Poulantzas 1976; Aglietta 1979; Bell 1982; Negri 1984; Harvey 2005; Marazzi 2010) – can contribute much to an understanding of the ways that political-economic crisis plays out, these more empirically or institutionally-founded accounts nonetheless generally leave aside the granular particulars of life in crisis. Some of these dynamics are touched upon by the 'cultural political economy' literature (Sum and Jessop 2013), but there the emphasis largely remains on the ways that political economy is 'culturalized' or culturally inflected. This is certainly a crucial insight, but what it does not illuminate are the ways in which, even within a given community or social group, the 'experience' of crisis can vary enormously even on relatively common 'cultural' terrain. The vastly differentiated impacts of mass home foreclosure across lines of race, class and gender (Wyly et al. 2009; Wyly and Ponder 2011), the grossly uneven distribution of the burdens of austerity or so-called 'fiscal consolidation' (Glasmaier and Lee-Chuvala 2011; Palaskas et al. 2015), or the specifics of national and regional institutions (Fishman 2012; Quaglia and Royo 2015) are not just superstructural details, but (as Marx might have put it) what crisis 'in the concrete' actually looks like.

In this respect, as the chapters that follow demonstrate, the financial crises since 2007–2008 have motivated an enormous diversity of exciting, insightful and innovative scholarship. The disciplinary, geographic and empirical range of this research defies summary, but arguably its most significant general characteristic is the analytical centrality of debt and credit. Across the social sciences, even in work that is not explicitly oriented to the financial crisis per se, studies of finance as a social relation and financialization as a mode of modern life have exploded.

In the early post-Lehman moments, many of these accounts were motivated by a journalistic drive to expose the 'greed' and impunity of those involved (Madrick 2012). But as time has passed, this understandable urge to name the guilty has receded as the 'greed' at the heart of banking, for example, is understood less and less as a function of bad individuals, and more as a product of the social relations of financialized capitalism (Lapavitsas et al. 2012; Brown 2015; Lazzarato 2015). As a result, the debtor–creditor relation appears in much greater complexity, constituted as it is not only in predatory theft and misrepresentation, but also in consensual, historically-embedded, and irreducibly social relations at multiple temporal and spatial scales (Chu 2010; Graeber 2011; Kear 2013; James 2015). In other words, scholars have increasingly come to see the crises not simply as the result of individual opportunism which produced unlucky or accidental outcomes in an otherwise stable institutional setting, but of systemic or quasi-systemic processes that are nonetheless irreducibly contingent in their historical and geographical specificity, caught up in times and spaces that are always but never only political-economic.

Meanwhile, if scholars have had the relative luxury of grappling with the financial crisis in the abstract and at their leisure, governments and their various regulatory agencies have clearly not been so fortunate. Their responses have been more urgent and, of course, significantly more material. And while sometimes explicitly guided by scholarly interpretations – as with the uses and abuses of Reinhart and Rogoff's (2010) research – their inspiration has often been considerably harder to discern. It is to the state's response that we now turn.

The Crisis, the State and Regulation: 'What have I done?'

Amidst the period of peak anxiety about the fate of the financial system during the global financial crisis of 2007–2009, it is interesting to speculate how many central bankers and finance ministers experienced what might be described as a 'Lieutenant Colonel Nicholson moment'. Nicholson is a character played by Alec Guinness in the 1957 motion picture, *The Bridge on the River Kwai* (Lean 1957), who commands British prisoners of war put to work constructing a railway bridge in Burma for their Japanese captors. Seeing the completion of the bridge as a way of building morale among his troops, Nicholson not only encourages his men to do a good job but intervenes to improve its location and design. However, given the military significance of the railway, Allied commandos are dispatched to destroy the bridge – an intervention that in his pride Nicholson initially attempts to thwart. Almost too late, he recognizes his misplaced allegiances with the tragic question 'What have I done?': his attempt to protect the bridge is an act of collaboration. Injured in a melee between the saboteurs and Japanese soldiers, Nicholson falls on an explosive detonator, destroying himself and his bridge.

While there have been no recorded acts of self-immolation by politicians, regulators or central bankers in the wake of the financial crisis, there is some evidence to suggest that at least some of those involved in the regulation of money and finance in the period leading up to the crisis belatedly realized that they, like Nicholson, may have been labouring under a fatal misapprehension of their role and purpose. Through acts of commission and omission they helped build a bridge that intensified the connections between the high risk, high stakes world of global finance and the everyday lives of millions of people (cf. Pryke and Allen 2000). Andrew Haldane, now chief economist of the Bank of England, has insisted that the crisis was the product of nothing less than the 'failure of an entire system of financial sector governance' (Haldane 2012, p. 21).

Seduced by the 'Great Moderation', a long period of global economic growth combined with low inflation in the global economy from the mid-1990s onwards, many elites convinced themselves this benign macroeconomic outcome was a result of submitting to the power and *ex post* organizing logic of markets. The neoliberal prophets who had long advocated this form of market organization had surely been proved right, and all was set fair now the former belief that the state could intervene to organize markets *ex ante* was recognized as both mistaken and futile. 'Meddling' would only create distortions and inefficiencies in the more beneficent, emergent 'spontaneous order' (Mirowski 2013; Peck 2010; Sayer 1995).

When the financial crisis broke, its unprecedented scale and scope soon became evident. Although it is impossible to know the precise number that would indicate the scale of economic loss, as Berry et al. (2016, p. 16) point out, even the lower limits are extraordinarily high for a financial crisis:

> Estimates of the GDP loss as a result of the crisis and ensuing recession, fuelled in part by the crunch in bank lending to productive firms, range from £1.8 trillion and £7.4 trillion (depending on how permanent the effects of the crisis were). New research by the Bank for International Settlements (BIS) suggests the figure is more likely to be higher than lower and that credit booms and busts inflict long-term damage on economies: 'First, credit booms tend to undermine productivity growth ... Second, the impact ... is much larger if a crisis follows.' Andrew Haldane has pointed out that the total loss of income and output caused by the banking crisis was equivalent to a World War.

Losses of this magnitude challenged the faith of some of the most devout believers in the self-organizing capacity of markets, including the high priest of central bankers, ex-Federal Reserve Chair Alan Greenspan. Long a vociferous advocate of neoliberal approaches to economic management, Greenspan had his own Nicholson moment in 2008 when, observing the damage to the financial system he had overseen for so long, he admitted he had 'discovered a flaw in the model that I perceived is the critical functioning structure that defines how the world works' (cited in Mann 2010, p. 601). Greenspan's ontological crisis was more practically observed under his successor, Ben Bernanke, who embraced neoliberal apostasy in the decision 'to block market verdicts on which banks should fail' (Mirowski 2014, p. 85), bailing out one failing institution after another in a process repeated across other leading financial economies.

The relationship between regulators and the regulated – that is, between the state and its agencies on the one hand, and financial institutions and their employees and customers on the other – is critical to understanding how financial systems evolve through time. The process of financial innovation has been described as 'bricolage' (Engelen et al. 2010), a process of 'creative tinkering' (MacKenzie 2003, p. 831) by financial institutions seeking ways around restrictions and limitations. As a result, financial innovation 'is contingent, resourceful and context-dependent' (Engelen et al. 2010, p. 54), ensuring the relationship between financial institutions and regulators is constantly in motion. This relationship is critical given the key role that money and finance play in the economy, and how financial instability and crises have serious implications for the integrity of states and their systems of government. Indeed, contemporary financial markets evolve through a continuous process of reregulation, adapting in response to the evolution of and innovation in markets and the disposition of central banks, regulators and states towards risk and uncertainty at different conjunctures.

At times, and particularly in the wake of major financial crises, the relationship between regulators and regulated is adversarial, the state imposing new rules and restrictions on financial actors, seeking to tame money and finance and bring them to the service of broader societal, economic and political strategic goals. Perhaps the best example of this is the period following the financial crisis of the late 1920s and the depression of the 1930s, which saw the US embark upon a wholesale redrafting of financial regulation to 'put finance in its place' (French and Leyshon 2012). In an attempt to prevent financial contagion in the event of a crisis, regulators constructed clear demarcations between different financial markets, lines which financial institutions could not cross. These 'regulatory innovations' included the Glass–Steagall provisions of the US Banking Act of 1933, which institutionally separated investment and retail banking operations. This was followed later by the development of the Bretton Woods system, which sought to impose similar regulatory capacity and oversight at the level of the international financial system in order to facilitate post-war recovery.

However, the relationship between regulators and the regulated can also be more collaborative and tactical, as politicians become convinced of the benefits of unshackled financial services, which might accrue in the form of contributions to economic output and growth, overseas earnings and employment and tax income of various kinds (Lee et al. 2009). This cooperative relationship between the state and financial markets has dominated since at least the early 1980s, facilitated at least partly by the active and influential lobbying of financial services firms (Engelen et al. 2011), and their ability to offer lucrative careers to politicians and regulators beyond government. Indeed, Johnson and Kwak (2010, p. 6) describe the effective capture of US government by financial interests, with 'profits and bonuses in the financial sector ... transmuted into political power through campaign contributions and the attraction of the revolving door'. The close links between the US Treasury Department and powerful financial interests were also identified as an influence on the heavily skewed disbursement of funds in the interests of banks by the Special Investigator General of the Troubled Asset Relief Program (TARP) that administered the US bailouts (Barofsky 2012), contrasting with TARP's far less diligently policed aim to 'preserve homeownership'. Nor is the revolving door between government and finance limited to the US.

For example, Tony Blair, Gordon Brown and Alistair Darling, accounting between them for every UK prime minister and/or chancellor of the exchequer in the 1997–2010 Labour government, were all offered lucrative post-government positions with leading financial institutions, and at the time of writing drew salaries from JP Morgan, Pimco and Morgan Stanley respectively (Pickard 2016). George Osborne, who succeeded Darling as chancellor following the 2010 general election, but who lost his job in the post-Brexit rise of Theresa May to prime minster, joined this exclusive club by subsequently accepting a job as an adviser to Blackrock, the world's largest fund manager (Treanor and Mason, 2017). For this reason the 'interests of the City of London' always figure strongly in UK debates about international treaties and accords, particularly with the European Union, despite the fact that any net economic gain from London's financial district needs to be set against the risk to the public finances from bailouts and a skewing of national economic development towards financial sector interests (Berry et al. 2016).

This is what made the result of the British referendum on the UK's membership in the European Union so shocking to the British banking industry, because the outcome of the vote to leave put membership of the single market in doubt. Since 1992, 'passporting' agreements have enabled financial services firms licensed in one EU member state to operate in any other, helping competitive and innovative UK firms gain market share across Europe. However, when the cost of the public bailout of the financial services industry was passed on to other parts of the British economy and society through a strategy of austerity realized in welfare cuts and reductions in public investment, it had particularly regressive impacts in those parts of the UK that missed out on the post-crisis economic 'recovery' (Gardiner et al. 2013). Many attribute the previously unimaginable election of Donald Trump to the US presidency in November 2016 to a similar sense of abandonment and alienation in so-called 'left-behind' sectors and places (Goodwin and Heath 2016; Coontz 2016; Stein 2016; Swain 2016).

Prior to these unprecedented political developments, the tight relationship between financial and government elites ensured the global financial crisis did not precipitate a return to the more restrictive regulatory mechanisms advocated by Keynesian economists – mechanisms like those, for example, that Keynes himself helped to introduce to the international policy arena during the creation of Bretton Woods. One reason for this, according to Helleiner (Helleiner 2010; Helleiner and Pagliari 2009), is simply that not enough time has passed. Writing in the immediate aftermath of the crisis, Helleiner argued it was inappropriate to compare the post-global financial crisis response to Bretton Woods because the latter had a long gestation during the economic and political turmoil of the 1930s and 1940s, and there were years of preparation for the 1944 meeting in New Hampshire. The various G20 meetings called in the immediate aftermath of the more recent crisis were too impromptu and too focused on crisis management to permit proper deliberation about long-term structural reform. Moreover, in addition to time, there is also the matter of context. As a result of markedly 'different political circumstances' between the two periods, 'those hoping for a "Bretton Woods moment" in the wake of the 2007–2008 crisis have had unrealistic expectations' (Helleiner 2010, p. 622):

> The Bretton Woods 'moment' took place well over a decade after the momentous financial crises of the early 1930s. The delay was not just a product of the unique historical circumstances of the era. It took time for old ideas and practices to lose their legitimacy and for new ones to emerge as models for the future. Even the constitutive phase was a time consuming process involving complicated and painstaking preparations over several years (and this despite the favourable circumstances mentioned earlier of concentrated power, shared social purpose and wartime conditions). (ibid., 624–5)

More than half a decade after Helleiner wrote these words, rather than 'old ideas and practices losing their legitimacy', there remains little appetite for serious debate about either the central role that money and finance play in contemporary economy and society, or the existential threat the existing financial system poses through the looming threat of another debilitating crisis. Rather, on both sides of the Atlantic, there seems to have been a turn to another political trope of the 1930s, namely casting around for scapegoats to blame for economic austerity, usually minority and immigrant populations. But in the financial sphere, the predominant approach is a continuation of the mode of regulation that has prevailed since the 1980s (Christophers 2016a): a combination of permissiveness in terms of market behaviour that meekly attempts to ensure solvency through checks on the capitalization of the balance sheets of banks. The initial trigger for this latter form of supervision was the so-called Less Developed Countries debt crisis of 1982, which led to the introduction of the Bank for International Settlements' (BIS) first Basel Accord on capital adequacy ratios. The Accord required banks to set aside eight per cent of the value of outstanding loans as provision against default. Yet, at the same time, wholesale and retail financial markets were being reregulated across North America, Europe and Asia to increase competition, with the purpose of delivering greater efficiencies, profitability and consumer 'freedom'. This included, in 1999, the final elimination of the Glass–Steagall provisions, which had been gradually eroded since the 1960s but were finally swept away when the Financial

Modernization Act removed the regulatory firewall between investment and retail banking markets, part of a wider freeing up of financial rules during the 1990s (Dymski 1999).

This double movement – seeking to control balance sheet risk on the one hand, by focusing attention on capital adequacy, while also reregulating existing financial markets and permitting new ones on the other – also impacted the orientation of financial business models, which became much more focused on fee income than traditional bank intermediation (Erturk and Solari 2007). As we discuss in more detail in the next section, fee income had the advantage, inter alia, of generating profits without having to set aside capital, while the process of securitization allowed banks and other lenders to move debt assets off their books by selling investors rights to the repayment income (MacDonald 1996a, 1996b; Leyshon and Thrift 2007; Wainwright 2009, 2012).

Of course, in retrospect, we can see that it was the build-up of many such layers of ever more complex and interlinked securitized debt that helped trigger the financial crisis. This occurred despite the continuing efforts to 'de-risk' the financial system by updating the BIS capital adequacy framework (Basel II), which from 2004 sought to introduce more sophisticated and nuanced risk control by making capital adequacy requirements more flexible and responsive to financial institutions' internal risk control models. In other words, the greater the Value at Risk (VaR) to which a bank was exposed, the more capital the bank needed to hold for security and solvency. Clearly, this was unsuccessful, given the size of the bailouts required following the crisis.

In the wake of the crisis, while there have been regulatory reforms that work towards reintroducing some of the structural regulations that formerly separated different kinds of financial markets, they are pale imitations of those of the 1930s and 1940s (Langley 2015). For example, the Financial Services (Banking Reform) Bill in the UK, the Liikanen Review in the EU, and the Dodd–Frank Wall Street Reform and Consumer Protection Act in the US (if it survives the Trump administration plans to repeal it) all seek to pull apart more speculative investment and securities business from retail banking's more mainstream and quotidian activities (Erturk 2016). But they stop well short of the structural and institutional separation of investment and retail banking that was a feature of the US financial system for over fifty years. At the same time, the BIS has introduced Basel III, which 'aims to re-capitalise banks by improving the earlier Basel II framework by (i) a set of adjustments to the risk algorithm for loss absorbing capital, the minimum capital required against risk and the definition of what qualifies for capital and

(ii) introducing a new framework to measure and manage the liquidity risk that arises from pre-crisis over-reliance of banks on short-term volatile money market funds' (Erturk 2016, pp. 1–2).

These regulatory efforts to ring fence different parts of the banking system aim to ensure that in the event of another major financial crisis, banks will be saved by *bail-in* rather than bail-out; that is, regulators aim to avoid the moral hazard problems associated with 'too big to fail' by ensuring that the balance sheets of stricken banks are bolstered by their investors, and not by states and tax payers (Eichengreen and Rühl 2001). The 'just deserts' quality of this approach is appealing, but it is not unproblematic. For instance, Berry et al. (2016) are critical of the bail-in regulations because post-crisis attempts to separate wholesale and retail banking have not been comprehensive, so that the spread of problems from one area to the other in the event of a crisis remains highly likely. Moreover, such measures are almost certainly unlikely to provide the volumes of capital that will be required in the event of a very large and significant crisis. As Berry et al. (2016, p. 9) argue, 'if "bail-in bonds" are predominantly held by pension funds, insurance companies and other financial institutions, bank losses will still ultimately be borne by ordinary citizens', while 'if losses exceed the amount of equity and bail-in-able debt the losses will still have to be borne by someone – either depositors or the state. In both cases politicians may conclude that the costs to the wider economy are unacceptable and offer bail-out rather than bail-in.' Furthermore, compared to the early post-crisis years in both the UK and the US there has been a sea change in the prevailing attitude to banking regulation, with the focus moving back to classifying the banking sector as an economic boon rather than a burden, so that regulation has become more accommodating (Berry et al. 2016; Eichengreen 2015).

For Berry et al. (2016), the clearest signal of this accommodation within the UK economy was a statement by the Bank of England, in late 2015, that the 'post-crisis period' was over. This declaration was accompanied by 'a string of concessions to big banks' (p. 5). These concessions included:

> Changes to the bank levy which benefit large international banks such as HSBC at the expense of smaller challenger banks; the sacking of Martin Wheatley as Chief Executive of the Financial Conduct Authority (FCA), a move which has been followed by a number of inquiries and investigations being dropped; a watered-down set of proposals for implementing the ring fence between retail and investment banking, particularly in relation to economic links with the rest of the group; a disappointingly weak report from the Competition and Markets Authority, which rules out action to

break up big banks and instead focuses on consumer switching behaviour; confirmation by the Bank of England that banks will not be asked to hold significantly more capital; imposing a time-limit on claims relating to mis-selling of payment protection insurance (PPI). (Berry et al. 2016, p. 5)

Along similar lines, Erturk (2016) argues that efforts to control default risk through some form of structural regulation involving improved capital adequacy requirements would have been insufficient on their own in any case. Driven by the demand for returns on capital to satisfy investors, yet feeling constrained by capital adequacy requirements, financial institutions have become ever more resourceful in the search for profits. Erturk draws attention to two 'post-global financial crisis' crises in markets beyond the focus of recent regulatory reform, but where the chasing of returns in ostensibly mundane markets – cash management and retail insurance – led to significant losses for leading US and UK banks (Erturk 2016). Here the risks go the other way, as it is retail market banking – in the form of poor cash management and compensation payments for mis-sold payment protection insurance – and not investment banking that is the source of crisis. A better solution, for Erturk (2016, p. 11), is to make retail banking closer to a utility which 'needs to be ring-fenced from the unrealistic return on equity expectations of stock markets'.

Erturk's analysis is notable for paying close attention to a subject often curiously absent or underplayed in discussion of post-crisis bank regulation: the actual details of banks' business models and of how these are (or are not) changing in the wake of the crisis and the governmental-regulatory response. In the context of the wider financial sector (i.e., beyond the big banks at the heart of the crisis), these changes are the focus of the third and final part of our critical post-crisis stock taking.

The Crisis and the Financial Sector: 'What shall we do now?'

Where the financial sectors (US and European) most immediately implicated in the financial crisis are concerned, it has frequently been said in the years since the heat of the crisis that little has fundamentally changed – lessons have not been learned, practices have not been transformed – but on close inspection it is clear this is simply not the case. To be sure, both the European and US financial sectors have strenuously resisted significant components of states' and regulators' reform agendas, often with great success; equally certainly, those agendas have been weaker than many observers had hoped. Consequently, changes undertaken by financial institutions have undoubtedly been *less* material than they might otherwise have been. Nevertheless, the financial sector *has* responded to the crisis with change, and not only under political-regulatory duress. Institutions, practices and markets now all look markedly different than they did. In the following pages we offer a brief critical assessment of some of the most striking and significant responses and transformations. We do so through the lens of an attribute that has been much invoked where finance, crisis and reform are in question: the property of *risk*. We suggest that the financial sector in general – and major transnational banks in particular – have responded to the crisis by way of substantial strategic reorientation *to* risk, resulting in considerable reconfiguration of overall financial-risk landscapes.

Such reorientation and reconfiguration have resulted in part from the 'derisking' obliged by key regulatory developments, of which Basel III is probably the most important. Although its implementation has varied geographically and remains far from complete, its three central, closely-connected principles – touched upon in the previous section – concern capital adequacy (the amount of equity capital banks must hold against assets weighted according to their perceived riskiness, so-called 'risk-weighted assets'), leverage (equity capital adequacy as simply measured against total assets), and liquidity (the liquid assets banks must hold to ensure they are able to meet short-term liabilities *and* the stable long-term borrowings they need in order to fund illiquid, long-term assets).

Banks have clearly reacted, although often they have done so more in anticipation of these new regulatory requirements than in response to their actual implementation, which has been widely delayed. Before the crisis, return on equity (ROE) had been financial institutions' and their executives' critical benchmark, the metric they targeted and according to which many of the latter were remunerated: leverage inevitably, and famously, rocketed since operating on wafer-thin equity-capital bases boosted returns. But the new capital adequacy requirements and long-term liquidity provisions encourage both a bolstering of equity and a simultaneous pruning of illiquid, long-term assets, which is to say those assets considered most risky (and previously considered essential to high ROE): things like loans to small businesses or subprime mortgages. A new mantra has therefore increasingly taken ROE's place: return on risk-weighted assets (RWA).

The upshot has been a widespread reconfiguration of bank balance sheets. Whereas before the crisis banks sought to shrink equity through share buybacks and the like, they have since focused on shrinking RWA and RWA-intensive business activities, with bond investment-and-trading a prime example. The Swiss bank UBS has emerged as the poster child of this new strategic paradigm, shedding traditionally-core divisions that fall foul of Basel III strictures (Noonan 2015). Meanwhile, and partly as a corollary, interest-bearing deposits have typically grown their share on both the assets and liabilities side of bank balance sheets (for example, Bailey, Bekker and Holmes 2015, p. 3–4). As assets (i.e., cash deposited *by* the bank), they help provide the short-term liquidity demanded by Basel III. As liabilities (cash deposited *with* the bank), they help provide the stable funding source also demanded by Basel III, unlike the shorter-term liabilities – such as the unsecured 'commercial paper' issued by corporations – they have partially supplanted.

Have such developments made banks and the financial system stronger and safer? The jury remains out, and with good reason, for there are meaningful grounds for caution in both regards. Where banks themselves are concerned, plenty of observers believe the aforementioned regulations have not gone nearly far enough in forcing balance sheet transformation and, therefore, neither have the banks (for example, Admati and Hellwig 2013). For most systemically-important banks, the ratio of liquid assets to total assets has increased only relatively modestly – for example, from 11 to 15 per cent for US banks and from 20 to 23 per cent for UK banks between 2009 and 2014 (Haldane 2015, p. 6) – and increases in ratios of equity capital to total assets, risk-weighted or otherwise, have tended to be equally modest (for example, Team 2016). The axiom that equity is more expensive than debt continues to hold sway.

The safety of the financial system more broadly is of course a much wider and even more complicated question. For one thing, there is the problem of the types of financial institutions that step into the breach when banks, for regulatory reasons, are encouraged to step back and vacate the market – given that the activities from which banks withdraw do not usually simply disappear. Kneejerk criticism of banks has of course been something of a default 'left' position since the crisis, and deservedly so, but such criticism is often problematically narrow. Too few commentators have asked what and who takes the place of the much-maligned banks. Could it sometimes be a case of better-the-devil-you-know?

Consider some implications of the capital adequacy rules. If banks can improve capital ratios by raising more equity, they can also do so simply by reducing lending and investment – and thus assets accrued – in areas deemed risky. Since the crisis, this is exactly what many have done. Among the assets accorded high risk weightings by regulators, and thus requiring disproportionate (i.e., expensive) equity cushions, are long-term loans to business. Hence it is no surprise that banks have cut back on such credit, potentially threatening economic growth prospects in an already precarious macroeconomic environment of feeble recovery from recession. The *Economist* (2014) reported that 'bank lending to businesses in America is still 6% below its 2008 high. In the euro zone, where it peaked in 2009, it has declined by 11%. In Britain it has plummeted by almost 30%.' But do such non-bank businesses simply cease attempting to borrow? Clearly not. They seek other sources, and the question of financial-system safety – not to mention customer protection – depends in part on what alternatives exist.

In the United Kingdom, as banks have withdrawn from lending to non-bank businesses, the latter have turned to two main alternative sources (ibid.). One is 'shadow banks' (non-bank financial intermediaries), which can lend more cheaply because they are not deemed banks nor regulated as such. Asset management companies have been prominent in this regard, and it is notable in a wider context that since the crisis the world's largest asset managers have become bigger, in terms of assets, than the world's largest banks (ibid.), prompting observers to wonder: 'The age of asset management?' (Haldane 2014). The second alternative source has been the bond markets – and again, not just in the United Kingdom: global corporate bond-issuance doubled between 2007 and 2012, to $1.7 trillion (*Economist* 2014). The combination of these two alternative sources – asset managers and bond markets – is writ large in the post-crisis growth of managed *bond funds*, the size of such funds in the United States, for example, more than doubling between 2009 and 2015, from $1.5 to $3.5 trillion (Haldane 2015, p. 6).

While we return to the significance of these bond funds in due course, here it bears asking what it might mean for financial-systemic safety as they supplement or even supplant banks as corporate funding sources in the post-crisis era. With cash and near-cash interest rates at historic lows, asset managers in search of yield have increasingly been investing in less liquid, more risky assets (i.e., precisely those that banks are abandoning). Haldane (2014, p. 3) reports that since 2008 high-yield bond funds have grown at an annual rate of around 40 per cent, 'outpacing growth in the global mutual fund industry by a factor of four.' Similarly, with the recent reemergence in the United Kingdom of the securitization of subprime mortgages as 'a major driver of the UK's overall securitisation market', it is notable that the main originators of these mortgages are not now the clearing banks but 'a new generation of lenders filling a void vacated by high street institutions' (Hale 2015). In short, there would appear to still be a great deal of risky investment taking place, only not by the banks. Does this mean we now have a safer financial system? It seems unlikely.

Furthermore, even if banks have improved their liquidity profiles since the crisis, this does not mean *markets* are more liquid (liquid markets being those facilitating rapid sale of an asset with little or no loss in its value). Bond and derivative markets have emerged as a particular source of concern in the post-crisis years. These primarily 'over-the-counter' markets are less formal, rules-based, centralized and transparent than 'exchange-based' (e.g., equity) markets. In particular, they rely heavily on the risk-taking, market-making capacity – which is to say, the liquidity provision – of major wholesale and investment banks; they rely on such banks standing ever-ready to take a buy or sell counter-position (profiting on the spread between the two). Since 2008, however, such banks have substantially withdrawn market-making capacity from these markets and liquidity has dried up, to the tune of as much as 35 per cent (measured in terms of market volumes as a percentage of amounts outstanding) according to some estimates (Redburn Research 2013, pp. 12–13).

This is clearly a function at least in part of regulatory changes: not only are banks now encouraged not to hold illiquid assets, especially long-dated bonds, under Basel III – and are hence less willing to do so – but Dodd–Frank's prohibition of proprietary trading at Federal Deposit Insurance Corporation-insured banks also steers them away from such assets. Indeed, in late 2015, the net inventory of corporate bonds held by the US Federal Reserve's twenty-two primary dealers (those banks permitted to trade directly with the Fed) fell below zero for the first time (Eddings 2015). Perhaps this is a good thing? Again, it seems unlikely. For all their sins, in the face of steep bond-market price declines the banks in question could in the past generally be relied upon to step in as dealers/buyers of last resort. Today, they are much more reluctant to do so, and with new 'substitute' capital intermediaries such as asset managers showing no signs of playing a similar role, it is left to the Fed to act as last-resort dealer (cf. Mehrling 2011).

Of course, banks' post-crisis wariness of risky asset classes is by no means entirely a function of regulation. Having been through the wringer with the subprime crisis – often finding that the risk they *thought* they had transferred to investors through securitization in fact remained on their balance sheets (Acharya, Schnabl and Suarez 2013) – banks are not keen to go through the process again. Once bitten, twice shy. Indeed, in case one were inclined to believe that Basel III and Dodd–Frank have served to frustrate banks chomping at the bit of risk assumption, banks' response to those aspects of the latter regulation that would have required them to bear substantive risk should give pause. Dodd–Frank's initial 'credit risk retention' proposal stipulated that at least 5 per cent of the risk of mortgages and other loans should 'stay behind' with the loan issuer (typically, the bank) when such assets were pooled and securitized: the banks should be required to have skin in the game. But the banks baulked at even this relatively modest proposal, lobbied hard against it, and eventually won the day (Norris 2013; Ashton and Christophers 2016).

Furthermore, even before the financial crisis, major western banks were already moving *out* of activities requiring significant exposure to market and balance-sheet risk, instead prioritizing those involving much less proprietary risk assumption (Erturk and Solari 2007, pp. 375–8). Fee-earning activities such as mergers and acquisitions advisory, capital-raising, and prime brokerage were pushed to the fore. The post-crisis period has not seen a move back 'into' risk bearing, but rather a demonstrable continuation and deepening of this particular shift in banks' favoured 'value models' (Christophers 2015b, pp. 8–9; cf. Noonan 2015). This shift, we should be clear, is highly strategic, and is not only a result of post-crisis regulatory reform, even if the latter has tended to reinforce it.

Alongside and in addition to this active retreat from risk and the refusal to bear it when instructed to, perhaps more importantly still, the banks in question have plainly been active in the post-crisis period in transferring risk to others – customers, investors and other counterparties. Some instances of such transfer have of course been egregious. The scandals in the setting of benchmark rates (e.g., LIBOR) in both foreign exchange and credit markets, exposed and prosecuted by investigators in 2013 through 2015, were explicitly about risk and the concerted and often collusive attempt by traders to load that risk onto others while minimizing or even mitigating proprietary risk altogether (Ashton and Christophers 2015). These manipulations represent behaviours about as far from the heroic mythology of the financial markets – outsized, manly rewards for outsized, manly risks in the face of intense competitive pressure – as it is possible to imagine. It was, inter alia, these manipulations that the International Monetary Fund (IMF) chief Christine Lagarde had in mind in 2014 when lamenting that the financial sector 'has not changed fundamentally in a number of dimensions since the crisis' (cited in Monaghan 2014).

But, significantly, it is certainly not only banks that are implicated in the post-crisis redistribution of financial risk – it is also other types of financial intermediary. And such redistribution is not always egregious. Indeed, it seldom is, and critical scrutiny of less overt mechanisms of risk transfer is arguably even more important, precisely because of its lower visibility and the likelihood that it therefore goes unexamined. Let us revisit in this regard the abovementioned bond funds, which, as we have seen, represent funds that accept money from multiple end-investors and invest it in bonds or other debt securities (rather than in, say, equities). We noted earlier the striking post-crisis growth of such funds and their signal exposure to risky asset classes from which banks have sought to distance themselves, like long-term, often high-yield, corporate debt. What we did not identify, however, was how they mediate risk on the liabilities side of the equation. Who is on the hook if things go awry, and what does it mean for system-wide stability?

This is an absolutely crucial issue. Asset managers that offer these funds are different *types* of intermediaries than banks, with markedly different risk exposures. 'Whereas a bank intermediates between savers and borrowers *by entering into separate transactions with each, with all the risk that entails*', explains the *Economist* (2014; emphasis added), the asset manager 'is merely a matchmaker, with no "skin in the game"'. In the specific case of today's UK bond funds, for example, end-investors give their money to their fund manager 'to lend to mid-sized British businesses. All the proceeds from the loans go to the investors, *who must also bear any losses*; [the manager] simply administers the portfolio of loans on their behalf and', naturally, 'charges them a fee' (ibid.). Unlike the bank, the manager does not bear risk; and thus in the post-crisis era not only, *pace* Haldane (2014, pp. 4–5), are managed funds 'increasingly being allocated into illiquid assets' (as we saw above), but also a 'progressively greater share of investment risk [is] being put back to end-investors, with commensurately less being borne by intermediaries and companies'.

So again, the question arises: what does this mean for post-crisis financial stability, safety and protection? The IMF (2015, p. 94) has recently had its say, arguing that 'the larger role of the asset management industry in intermediation has many benefits', particularly 'from a financial stability point of view'. How so? 'Banks are predominantly financed with short-term debt, exposing them to both solvency and liquidity risks. In contrast, most investment funds issue shares, and end investors bear all investment risk.' But of course, this is just as one-eyed as the same venerable institution's (in)famous celebration – in 2006, of all years – of securitization for helping 'make the banking and overall financial system more resilient' through 'the dispersion of credit risk by banks to a broader and more diverse group of investors' (IMF 2006, p. 51). In fact the parallels, at least to us, seem uncanny. As the altogether more astute Andrew Haldane (2014, p. 5), of the Bank of England, acknowledges, these post-crisis trends in 'the contractual structure of liabilities' have 'risk implications' not just for the financial system and its putative stability but also *specifically* for the end-investors that the IMF gaily sees 'bearing all' investment risk. 'A key question, here', Haldane (ibid.) correctly observes, 'is how household behaviour is likely to respond to bearing these additional risks, especially in situations of stress' (c.f. Berry et al. 2016). Haldane's answer – that it is likely to respond badly – is not very comforting.

One of the key insights of cultural-economy work on finance over the past decade or so, meanwhile, has been that the active transfer and redistribution (to others) of financial risk, epitomized by the asset management industry, entails and relies upon the creation and enrolment of new risk-bearing subjects (Maurer 1999): subjects who, under conditions not of their own choosing, come to feel some combination of capability, compulsion and/or craving to embrace the risk inherent in financial products. These processes of subject-formation were, of course, much in evidence before the crisis, and indeed were arguably essential to its materialization (Langley 2007; Kear 2014); but they have only intensified in the years since, as finance capital seeks to extend and expand its accumulative ambit by adding, at economics' storied 'margins', new, previously-excluded financial subjects and thus new risk-bearers to the ranks of 'homo subprimicus' (Kear 2013) and the like. Roy's (2010) analysis of microfinance and the new population of entrepreneurial debtors it has crystallized in the Global South highlights a relatively well-known and much-studied example. Less well known, but arguably just as important, is the largely post-crisis growth in the same parts of the world of index-insurance products that help render rural smallholders creditworthy – calculated as capable of bearing debt – in the first place (Johnson 2013).

We submit that like other notable features of the post-crisis landscape of money and finance, these new risk-bearing subjects – so crucial to finance capital's expansive imperative – can be helpfully and critically understood in terms of the conjuncture of three sets of closely interrelated phenomena. One is the set of transformative developments in modern-day capitalism often brought together under the umbrella concept of 'financialization': the intensification of financial logics and dynamics within realms where capitalist finance has long been present, if you like, alongside such finance's colonization of hitherto non-financialized domains (as with microfinance and index insurance). The second phenomenon comprises the actual substantive financial practices that, amongst other things, effect financialization and give it purchase. Critical scholarship on money and finance cannot eschew engagement with the nitty-gritty of how, in practice, they 'work' (Ho 2009; Luyendijk 2015); if it does not understand practices, it veers more or less inevitably towards acritical abstraction. What is the business model of microfinance? Of index insurance? What exactly do they require of their customers, and how do they configure and apportion risk? We need to know. And third, we need to recognize the power of ideas and imaginaries. The political and cultural economy of capitalist society is indelibly shaped by the knowledges it generates and that circulate within it – no more so, we suggest, than in the context of money and finance (Christophers 2013). These imaginaries include those of the 'defunct economists' famously highlighted by Keynes (1965, p. 383), but are not limited to them. Indeed, if the chapters that follow tend to focus somewhat more on the realm of finance than on that of the monetary in general, it is perhaps a product of these irreducibly social imaginaries, since for most of us, the realm of money comes to life in the category of finance. In sum, we urge critical scholarship on post-crisis money and finance at the intersection of financialization, financial practices and financial imaginaries. The contributions in this book fulsomely demonstrate the significance of each of these themes.

The Content of the Book

The remaining chapters in this book are organized into the three broad aforementioned categories. First, there are those chapters that address *financial imaginaries* of one kind or another. For example, Dick Bryan, Michael Rafferty and Duncan Wigan explore the ways in which financial techniques and practices have evolved and developed in relation to the power and authority of the state, and how such techniques and practices have challenged established spatial framings and understandings of the national interest and appropriate jurisdiction. Bryan, Rafferty and Wigan focus on struggles to define the appropriate spatial scale for both understanding and governing money and finance. Many accounts of the development of the global financial system see it as a process that involves a shift from a constellation of inter-connected but internally coherent national financial economies to one that now better resembles a space of flows, wherein the power of financial capital has usurped that of the state. However, Bryan, Rafferty and Wigan argue that this linear account that moves from the national to the global scale is too simplistic, and that it is necessary to identify an 'in-between' financial imaginary, where both the mobility of capital and the residual obduracy and fiscal base of the state matters. Traditional means of national accounting and framing have indeed been undermined and subverted by financial innovation, and by the development in particular of financial derivatives. The ways in which financial derivatives allow capital to bend and distort established understandings of where financial activity takes place has helped to undermine the concept of a national economy. Nevertheless, geographically-distinctive economies, backed up by the power of the state, remain important to global finance. For example, the existence of discontinuous juridical spaces in the global economy provides ample opportunities for financial institutions to engage in regulatory arbitrage, which in turn expedites the reregulation of financial markets in favour of mobile and fungible capital. Moreover, as the world's most state-subsidized economic activity, finance remains reliant on nation states: private losses were socialized during the bail-outs of the financial crisis and less mobile citizens were called upon to recapitalize financial institutions through national tax systems.

Meanwhile, Paul Langley undertakes a critical analysis of the ways in which spatial imaginaries underlie the speculative circulations of financial capitalism, and how the mobility and freedom of capital has been justified as 'vital' to economic well-being. Langley reveals the metaphors that underpin understandings of the contemporary financial system, and how these are used to justify an order that supports and validates the unhindered circulation of money. Thus, the breaking of the financial crisis was a threat to 'liquidity' and to the ability of money and finance to flow where it was 'needed', or at least where it could turn a profit, without which the wheels of the economy would, it was claimed, seize up. Within this spatial imaginary, it was imperative that the speculative circulations of finance be restored, justifying state intervention that would take responsibility for nonperforming assets and repair the balance sheets of banks and other financial institutions so that capital could flow freely once more. But, as Langley points out, in the management of the crisis there was very little consideration of alternative spatial imaginaries that might have looked to constrain the speculative tendencies of capital, as happened in the wake of the financial crisis of the 1920s and 1930s. This, he suggests, may be indicative of how financial elites had managed to successfully place certain aspects of the financial system beyond political scrutiny even in the midst of a catastrophic crisis.

One of the reasons for this, Gary Dymski suggests, is the ways in which orthodox economics has become so powerful in framing economic policy. Dymski argues that one way of challenging this dominance and economics' rather simplistic and reductionist spatial imaginary would be for geographical research into money and finance to properly incorporate concepts of financial instability into their accounts. Despite geographers' extensive writings on crises of various kinds, Dymski identifies the lack of engagement with economics as a weakness, in as much as it prevents economic geography making stronger analytical connections with otherwise likeminded heterodox economists, and particularly those that seek to develop the ideas of Keynes and followers such as Minsky. Given the importance of orthodox economic ideas in the development and formulation of policy, Dymski argues that notwithstanding geography's spatial insights, unless Keynesian ideas are brought into the geographies of money and finance this body of research will lack the conceptual firepower to effectively counter the practices and policies that lead to endemic instability in the first place and then the production of austerity as a remedial counter measure. Although this call is a challenge, Dymski also claims it is an opportunity because by engaging with Keynesian ideas geographers would bring a greater spatial awareness to an allied non-orthodox account and so be better able to challenge the geographical blindness inherent within orthodox economic accounts.

Second, there are those chapters that broadly deal with the conduct of *financial practices* of various kinds. Marieke de Goede explores the emergence of what she describes as a 'finance-security assemblage', and the ways in which financial institutions can be called to account by states – and by the United States in particular – for actions that may be interpreted as enabling terrorist activity. In so doing, de Goede also draws attention to the power exerted by the US in bringing financial institutions into line through punitive sanctions. Her chapter connects to the arguments in previous chapters, in that to counter the perceived threat from terrorist financing, a new spatial imaginary was required; and here the US sees its jurisdictional power as transactional, which extends to those places where exchanges in dollars take place, rather than being territorially limited to domestic space. This transactional lens gives the US global reach in such matters, and has had material consequences for the practices of banks in regards to compliance and the calculation of risk. As de Goede illustrates, such is the fear of the large penalties that can be imposed for facilitating the financing of terrorism that many financial institutions have adopted a highly conservative approach to lending where transactions may be 'red flagged' by a combination of software and professional judgement. As a result, some international banks have chosen to deny service to some customers associated with Muslim faith groups and charities, even when there is no material evidence of a link with terrorist activity. This process of anticipatory 'derisking' is having equity effects, particularly when account closures have interfered with the flow of remittances to impoverished parts of the world that have come to rely on them for development needs.

Unintended equity effects are also addressed in Deborah James' chapter, which focuses on efforts to address the 'credit apartheid' that emerged during the period of white minority rule in South Africa. What James illustrates is that while states may wish to enact reregulation to bring about social and economic change, such intentions are often subverted by the intervention and practices of intermediaries that operate in the institutional and market spaces that are the target of such regulations. The impoverishment of black South Africans has its roots in the apartheid regime and James outlines how a long history of low wages and exploitation enabled the credit industry – both licensed and unlicensed – to incorporate itself into the lives of poor communities by offering, for a price, a level of consumption that would otherwise have been impossible. However, because blacks were denied access to property, creditors sought ways to attempt to 'secure' unsecured debt, which saw the state intervene to enable creditors to have access to workers' salaries so that loans could be repaid at source. But while the overhaul of this iniquitous credit system was a priority following the introduction of democracy in the 1990s, in practice the eradication of exploitative credit relations has proved problematic. For one thing, a democratic political system saw a democratization of access to credit as firms flooded into the market to provide credit to a population that was enthusiastic and optimistic about social and economic progression. But this process of financial inclusion increased problems of indebtedness. Indeed, as James reveals, attempts to reform the credit system to make it more responsible have been undermined by both creditors and debtors: by the former in their desire to make money, and by the latter in their desire to obtain money to consume in an environment of rising but still low incomes. Indeed, some borrowers used their loans to set up new micro lending businesses through which they could pass on their debts, at a profit, both deepening and extending networks of credit and debt.

The third set of chapters focuses on processes of *financialization*, broadly defined. Phillip O'Neill looks at the ways in which the building of urban infrastructure, long a preserve of the state and paid for by public funding, has increasingly been financed by private capital as a way of building public goods but without recourse to public sector financing. The requirements for managers of large pools of financial capital to locate long-term assets to generate returns has seen investment funds pay for a wide range of infrastructure assets so that rents can be extracted from their use. While this shift from public to private sector funding has permitted the building of infrastructure that may not otherwise have taken place within states averse to public sector borrowing and tax increases, O'Neill argues that the financialization of infrastructure construction and the giving over of once public assets to private financial interests has implications both for our understanding of public goods and the expectations of positive socio-environmental externalities that are traditionally produced by urban infrastructure.

Jessica Dempsey's chapter, meanwhile, considers the financialization of nature. Its advocates claim that this process is a way of preventing biodiversity loss by placing the environment within an economic and financial calculative frame; the development of concepts such as ecosystems services and natural capital seeks to encourage investors to view the environment as a potential asset that is more valuable when preserved and exploited in a sustainable and non-disruptive manner. However, Dempsey argues that three connected sets of concerns and uncertainties continue to surround the financialization of biodiversity and critical perspectives on it. First, the production of guaranteed income streams from conservation financialization has in fact proved more elusive that many critics suggest. Second, it is therefore unclear what the effects of such financialization will be, if and when it materializes more substantively. Third – and perhaps most thornily for critics of financialization – might a robust flow of profit-seeking finance capital into conservation actually represent an improvement on the status quo, which is arguably one of too little conservation-oriented capital of *any* description?

Finally, Karen Lai and Joseph Daniels' chapter focuses on the state-sponsored financialization of Singaporean banks, which sought to make the financial system more resilient and better suited to competing within global financial markets and delivering developmental outcomes to Singapore. Regulatory reform ushered in processes of capital centralization and concentration so that Singaporean banks became bigger with higher levels of capitalization. However, as Lai and Daniels argue, making reference to the variegated capitalism literature, a uniform programme of regulatory change produced a complex financial ecology as financial services firms responded in different ways to the state's injunctions, which produced a set of distinctive global financial institutions. Moreover, although Singaporean banks were damaged in the global financial crisis, they were not put in peril as was the case in Europe and North America. As Lai and Daniels argue, this is an important reminder that the global financial crisis did not necessarily carry the same historical and political significance in Asia as it did in the North Atlantic.

In sum, the chapters that follow analyse a wide range of 'post-crisis' monetary and financial issues from a wide range of critical perspectives. They do not cover the entire spectrum of money and finance, by any means; but they do demonstrate the centrality of money and finance to contemporary capitalism and its political and cultural economies. Nor do the chapters approach money and finance from a unified 'critical' perspective; but they do show that a critical perspective of some sort is essential to grappling with the capitalist present in its proliferating monetary and financial incarnations. We hope, as such, that the chapters can serve as something of a touchstone or lightning rod for similarly-inclined critical analysis in the years ahead.

References

Acharya, V., P. Schnabl, and G. Suarez. 2013. Securitization without risk transfer. *Journal of Financial Economics* 107:515–36.

Adjoran, I. 2014. The fetish of finance: metatheoretical reflections on understandings of the 2008 crisis. *Critical Historical Studies* 1:285–313.

Admati, A., and M. Hellwig. 2013. *The Bankers' New Clothes: What's Wrong with Banking and What to Do About It.* Princeton, NJ: Princeton University Press.

Aglietta, M. 1979. *A Theory of Capitalist regulation: The US Experience*. London: New Left Books.

Ashton, P. 2011. The financial exception and the reconfiguration of credit risk in US mortgage markets. *Environment and Planning A* 43 (8):1796–812.

Ashton, P., and B. Christophers. 2015. On arbitration, arbitrage and arbitrariness in financial markets and their governance: Unpacking LIBOR and the LIBOR scandal. *Economy and Society* 44 (44):188–217.

Ashton, P., and B. Christophers. 2016. Remaking mortgage markets by remaking mortgages: US housing finance after the crisis. *Economic Geography* doi: 10.1080/00130095.2016.1229125.

Backhouse, R., and B. Bateman. 2011. *Capitalist Revolutionary: John Maynard Keynes*. Cambridge: Cambridge University Press.

Bailey, M., W. Bekker, and S. Holmes. 2015. The big four banks: The evolution of the financial sector, Part I, https://www.brookings.edu/wp-content/uploads/2016/06/big_four_banks_evolution_financial_sector_pt1_final.pdf.

Barofsky, N. 2012. *Bailout: An Inside Account of How Washington Abandoned Main Street While Rescuing Wall Street*. New York: Free Press.

Bell P., and H. Cleaver. 1982. Marx's crisis theory as a theory of class struggle. *Research in Political Economy* 5:189–261.

Berry, C., D. Lindo, and J. Ryan-Collins. 2016. Our friends in the City: Why banking's return to business as usual threatens our economy. London: New Economics Foundation.

Blume, L.E., and S.N. Durlauf. 2016. Capital in the twenty-first century: a review essay. *Journal of Political Economy* 123:749–77.

Brown, W. 2015. *Undoing the Demos: Neoliberalism's Stealth Revolution*. Brooklyn, NY: Zone Books.

Carson, A. 2016. What to say of entirety? *New York Review of Books*:25 February, p. 4.

Chick, V., and S. Dow. 2012. On causes and outcomes of the European crisis: ideas, institutions and reality. *Contributions to Political Economy* 31:51–66.

Christophers, B. 2013. *Banking across Boundaries: Placing Finance in Capitalism*. Oxford: Wiley-Blackwell.

Christophers, B. 2015a. Geographies of finance II: Crisis, space and political-economic transformation. *Progress in Human Geography* 39 (2):205–13.

Christophers, B. 2015b. Value models: finance, risk, and political economy. *Finance and Society* 1 (2):1–22.

Christophers, B. 2016a. Geographies of finance III: Regulation and 'after-crisis' financial futures. *Progress in Human Geography* 40 (1):138–48.

Christophers, B. 2016b. *The Great Leveler: Capitalism and Competition in the Court of Law*. Cambridge, MA: Harvard University Press.

Christophers, B. 2017. The performativity of the yield curve. *Journal of Cultural Economy* 10 (1):63–80.

Chu, J. 2010. *Cosmologies of Credit: Transnational Mobility and the Politics of Destination in China*. Durham, NC: Duke University Press.

Coontz, S. 2016. Why the white working class ditched Clinton. CNN: 11 November, http://www.cnn.com/2016/11/10/opinions/how-clinton-lost-the-working-class-coontz/.

DeLong, J.B., and L.H. Summers. 2012. Fiscal policy in a depressed economy. *Brookings Papers in Economic Activity* Spring:233–74.

Duménil, G., and D. Lévy. 2011. *The Crisis of Neoliberalism*. Cambridge, MA: Harvard University Press.

Dymski, G.A. 1999. *The Bank Merger Wave: The Economic Causes and Social Consequences of Financial Consolidation*. Armonk, New York: M.E. Sharpe.

Eatwell, J., and M. Milgate. 2011. *The Fall and Rise of Keynesian Economics*. Oxford: Oxford University Press.

Eddings, C. 2015. Goldman contrarian joins chorus warning on bond-market liquidity. *Bloomberg Business*: 10 November, https://www.bloomberg.com/news/articles/2015-11-10/goldman-contrarian-joins-chorus-warning-on-bond-market-liquidity.

Eichengreen, B. 2015. Financial crisis: revisiting the banking rules that died by a thousand small cuts. *Fortune*:16 January 2015: http://fortune.com/2015/01/16/financial-crisis-bank-regulation/.

Eichengreen, B., and C. Rühl. 2001. The bail-in problem: systematic goals, ad hoc means. *Economic Systems* 25 (1):3–32.

Engelen, E., I. Erturk, J. Froud, S. Johal, A. Leaver, M. Moran, A. Nilsson and K. Williams. 2011. *After the Great Complacence: Financial Crisis and the Politics of Reform*. Oxford: Oxford University Press.

Engelen, E., I. Erturk, J. Froud, A. Leaver and K. Williams. 2010. Reconceptualizing financial innovation: frame, conjuncture and bricolage. *Economy and Society* 39 (1):33–63.

Erturk, I. 2016. Financialization, bank business models and the limits of post-crisis bank regulation. *Journal of Bank Regulation* 17 (1):60–72.

Erturk, I., and S. Solari. 2007. Banks as continuous reinvention. *New Political Economy* 23 (3):369–83.

Fishman, R.M. 2012. Anomalies of Spain's economy and economic policy-making. *Contributions to Political Economy* 31:67–76.

Foster, J.B., and H. Magdoff. 2009. *The Great Financial Crisis: Causes and Consequences*. New York, NY: Monthly Review Press.

Fourcade, M., E. Ollion and Y. Algan. 2015. The superiority of economists. *Journal of Economic Perspectives* 29 (1):89–114.

Fraser, N. 2013. *Fortunes of Feminism: From State-managed capitalism to Neoliberal crisis*. London: Verso.

Fraser, N. 2015. Legitimation crisis? On the political contradictions of financialized capitalism. *Critical Historical Studies* 2 (2):157–89.

French, S., and A. Leyshon. 2012. Dead pledges: mortgaging time and space. In *The Oxford Handbook of the Sociology of Finance*, eds K. Knorr Cetina and A. Preda, 357–75. Oxford: Oxford University Press.

Friedman, B. 2007. What we still don't know about monetary and fiscal policy. *Brookings Papers in Economic Activity* Fall:49–71.

Gardiner, B., R. Martin, P. Sunley and P. Tyler. 2013. Spatially unbalanced growth in the British economy. *Journal of Economic Geography* 13 (6):889–928.

Glasmaier, A.K., and C.R. Lee-Chuvala. 2011. Austerity in America: gender and community consequences of restructuring the public sector. *Cambridge Journal of Regions, Economy and Society* 4:457–74.

Gongloff, M. 2013. Reinhart, Rogoff backing furiously away from austerity movement. *Huffington Post*: 3 May, http://www.huffingtonpost.com/2013/05/02/reinhart-rogoff-austerity_n_3201453.html.

Goodwin, M.J., and O. Heath. 2016. The 2016 referendum, Brexit and the left behind: an aggregate-level analysis of the result. *The Political Quarterly* 87 (3):323–32.

Gorton, G., and A. Metrick. 2012. Getting up to speed on the financial crisis: a one-weekend-reader's guide. *Journal of Economic Literature* 50:128–50.

Graeber, D. 2011. *Debt: The First 5000 Years*. Brooklyn, NJ: Melville House.

Haldane, A. 2012. The doom loop. *London Review of Books* 34 (5):21–2.

Haldane, A. 2014. The age of asset management? Speech given by Andrew G. Haldane, Executive Director, Financial Stability and member of the Financial Policy Committee, 4 April, http://www.bankofengland.co.uk/publications/Documents/speeches/2014/speech723.pdf.

Haldane, A. 2015. Stuck. Speech given by Andrew G. Haldane, Chief Economist, Bank of England, 30 June, http://www.bankofengland.co.uk/publications/Documents/speeches/2015/speech828.pdf.

Hale, T. 2015. Bundling of risky UK mortgages booms. *Financial Times*: 8 October, http://www.ft.com/intl/cms/s/0/88bc8cfa-6ce4-11e5-8171-ba1968cf791a.html#axzz3yLFugRWF.

Hansen, K.B. 2015. Contrarian investment philosophy in the American stock market: on investment advice and the crowd conundrum. *Economy and Society* 44 (4):616–38.

Harvey, D. 2005. *A Brief History of Neoliberalism*. Oxford: Oxford University Press.

Harvey, D. 2010. *The Enigma of Capital and the Crises of Capitalism*. London: Profile.

Harvey, D. 2014. *Seventeen Contradictions and the End of Capitalism*. Oxford: Oxford University Press.

Helleiner, E. 2010. A Bretton Woods moment? The 2007–2008 crisis and the future of global finance. *International Affairs* 86 (3):619–36.

Helleiner, E., and S. Pagliari. 2009. Towards a new Bretton Woods? The first G20 leaders summit and the regulation of global finance. *New Political Economy* 14 (2):275–87.

Herndon, T., M. Ash, and M. Pollin. 2014. Does high public debt consistently stifle economic growth? A critique of Reinhart and Rogoff. *Cambridge Journal of Economics* 38:257–79.

Ho, K. 2009. *Liquidated: An Ethnography of Wall Street*. Durham NC and London: Duke University Press.

IMF. 2006. Global financial stability report April 2006: market developments and issues. https://www.imf.org/external/pubs/ft/GFSR/2006/01/index.htm.

IMF. 2015. Global financial stability report April 2015: navigating monetary policy challenges and managing risks. http://www.imf.org/External/Pubs/FT/GFSR/2015/01/pdf/text.pdf.

James, D. 2015. *Money From Nothing: Indebtedness and Aspiration in South Africa*. Palo Alto, CA: Stanford University Press.

Johnson, L. 2013. Index insurance and the articulation of risk-bearing subjects. *Environment and Planning A* 45:2663–681.

Johnson, S., and J. Kwak. 2010. *13 Bankers: The Wall Street Takeover and the Next Financial Meltdown*. New York: Pantheon.

Joseph, M. 2014. *Debt to Society: Accounting for Life Under Capitalism*. Minneapolis, MN: University of Minnesota Press.

Kay, J. 2011. The map is not the territory: an essay on the state of modern economics. Institute for New Economic Thinking, 4 October. https://www.johnkay.com/2011/10/04/the-map-is-not-the-territory-an-essay-on-the-state-of-economics.

Kear, M. 2013. Governing homo subprimicus: beyond financial citizenship, exclusion, and rights. *Antipode* 45 (4):926–46.

Kear, M. 2014. The scale effects of financialization: The Fair Credit Reporting Act and the production of financial space and subjects. *Geoforum* 57:99–109.

Keynes, J.M. 1965. *The General Theory of Employment, Interest and Money*. London: Macmillan.

Knight, F.H. 1921. *Risk, Uncertainty and Profit*. Boston, MA: Houghton Mifflin.

Kosselleck, R. 2006. Crisis. Trans.M. Richter. *Journal of the History of Ideas* 62 (2):357–400.

Krippner, G.R. 2011. *Capitalizing on Crisis: The Political Origins of the Rise of Finance*. Cambridge, MA: Harvard University Press.

Krugman, P. 2012. *End this Depression Now!*. New York, NY: W.W. Norton.

Krugman, P. 2014. Why we're in a new gilded age. *New York Review of Books*: 8 May.

Krusell, P., and A.A. Smith. 2015. Is Piketty's 'second law of capitalism' fundamental? *Journal of Political Economy* 123:725–48.

Kumar, M., and J. Woo. 2010. Public debt and growth. International Monetary Fund Working Paper WP/10/174.

Langley, P. 2007. Uncertain subjects of Anglo-American financialization. *Cultural Critique* 65:67–91.

Langley, P. 2015. *Liquidity Lost*. Oxford: Oxford University Press.

Lapavitsas, C., A. Kaltenbrunner, G. Labrinidis, D. Lindo, J. Meadway, J. Michell, J.P. Painceira, E. Pires, J. Powell, A. Stenfors, N. Teles and L. Vatikiotis. 2012. *Crisis in the Eurozone*. London: Verso.

Lazzarato, M. 2015. *Governing by Debt*. South Pasadena, CA: Semiotext(e).

Lean, D. 1957. *The Bridge on the River Kwai*. Los Angeles: Columbia Pictures.

Lee, R., G.L. Clark, J. Pollard and A. Leyshon. 2009. The remit of financial geography: before and after the crisis. *Journal of Economic Geography* 9 (5):723–47.

Leyshon, A., and N. Thrift. 2007. The capitalization of almost everything: the future of finance and capitalism. *Theory, Culture & Society* 24:97–115.

Lo, A.W. 2012. Reading about the financial crisis: a twenty-one book review. *Journal of Economic Literature* 50:151–78.

Lucas, R. 2003. Macroeconomic priorities. *American Economic Review* 93:1–14.

Lucas, R. 2009. In defense of the dismal science. *The Economist*: 6 August.

Luyendijk, J. 2015. *Swimming With Sharks: My Journey into the World of the Bankers*. London: Guardian Faber.

MacDonald, H. 1996a. Expanding access to the secondary mortgage markets: the role of central city lending goals. *Growth and Change* 27 (3):298–312.

MacDonald, H. 1996b. The rise of mortgage-backed securities: Struggles to reshape access to credit in the USA. *Environment and Planning A* 28 (7):1179–98.

MacKenzie, D. 2003. An equation and its worlds: bricolage, exemplars, disunity and performativity in financial economics. *Social Studies of Science* 33 (6):831–68.

Madrick, J. 2012. *Age of Greed: The Triumph of Finance and the Decline of America, 1970 to the Present*. New York, NY: Vintage.

Mankiw, G. 2015. Yes, r > g. So what?'. *American Economic Review* 105 (5):43–7.

Mann, G. 2010. Hobbes' redoubt? Toward a geography of monetary policy. *Progress in Human Geography* 34 (5):601–25.

Mann, G. 2015. A General theory for our time: on Piketty. *Historical Materialism* 23:106–40.

Mann, G. 2017. *In the Long Run We are All Dead: Keynesianism, Political Economy and Revolution*. London: Verso.

Marazzi, C. 2010. *The Violence of Financial Capitalism*. Cambridge: MIT Press.

Maurer, B. 1999. Forget Locke? From proprietor to risk-bearer in new logics of finance. In *Public Culture*, 47–67: Duke University Press.

McNally, D. 2001. *Global Slump: The Economics and Politics of Crisis and Resistance*. Oakland, CA: PM Press.

Mehrling, P. 2011. *The New Lombard Street: How the Fed Became the Dealer of Last Resort*. Princeton, NJ: Princeton University Press.

Mirowski, P. 2010. The great mortification: economists' responses to the crisis of 2007: (and counting). *The Hedgehog Review* 12 (2): http://www.iasc-culture.org/THR/THR_article_2010_Summer_Mirowski.php.

Mirowski, P. 2013. *Never Let a Serious Crisis go to Waste: How Neoliberalism Survived the Financial Meltdown*. London: Verso.

Monaghan, A. 2014. IMF chief says banks haven't changed since financial crisis. *Guardian*: 27 May, http://www.theguardian.com/business/2014/may/27/imf-chief-lagarde-bankers-ethics-risks.

Moore, J.W. 2015. *Capitalism in the Web of Life: Ecology and the Accumulation of Capital*. London: Verso.

Negri, A. 1989. *Marx beyond Marx: Lessons on the* Grundrisse. Baltimore: AK Press.

Noonan, L. 2015. Regulatory changes force investment banks into 'capital light' activities. *Financial Times*:13 December, http://www.ft.com/cms/s/0/566ed97e-8a2b-11e5-90de-f44762bf9896.html.

Norris, F. 2013. Mortgages without risk, at least for the banks. *The New York Times*: 28 November, http://www.nytimes.com/2013/11/29/business/mortgages-without-risk-at-least-for-the-banks.html?_r=0.

O'Connor, J. 1973. *The Fiscal Crisis of the State*. Piscataway NJ: Transaction.

Palaskas, T., Y. Psycharis, A. Rovolis and C. Stoforos. 2015. The asymmetrical impact of the economic crisis on unemployment and welfare in Greek urban economies. *Journal of Economic Geography* 15:973–1007.

Palley, T. 2011. Quantitative easing: a Keynesian critique. *Investigación Económica* 70 (2):69–86.

Panitch, L., and S. Gindin. 2013. *The Making of Global Capitalism: The Political Economy of American Empire*. London: Verso.

Peck, J. 2010. *Constructions of Neoliberal Reason*. Oxford: Oxford University Press.

Pickard, J. 2016. Labour figures swap politics for prominent business roles. *Financial Times*:2 February: http://www.ft.com/cms/s/0/3c1d0e58-c8f4-11e5-a8ef-ea66e967dd44.html#axzz40cAAKjED.

Piketty, T. 2014. *Capital in the Twenty-first Century*. Cambridge, MA: Belknap Press.

Posner, R. 2009. Liberals forgetting Keynes. *The Atlantic*:27 July.

Postone, M. 2012. Thinking the global crisis. *South Atlantic Quarterly* 111:227–49.

Poulantzas, N. 1976. *The Crisis of the Dictatorships: Portugal, Greece, Spain*. London: New Left Books.

Pryke, M., and J. Allen. 2000. Monetized time–space: derivatives – money's 'new imaginary'? *Economy and Society* 29 (2):264–84.

Quaglia, L., and S. Royo. 2015. Banks and the political economy of the sovereign debt crisis in Italy and Spain. *Review of International Political Economy* 22:485–507.

Redburn Research. 2013. Capital markets: the arteries are clogging. (Personal communication: copy available from authors on request.)

Reinhart, C.M., and K. Rogoff. 2009. *This Time is Different: Eight Centuries of Financial Folly*. Princeton, NJ: Princeton University Press.

Reinhart, C.M., and K. Rogoff. 2010. Growth in a time of debt. *American Economic Review* 100 (2):573–8.

Rodrik, D. 2015. *Economics Rules: The Rights and Wrongs of the Dismal Science*. New York, NY: Norton.

Roitman, J. 2014. *Anti-crisis*. Durham, NC: Duke University Press.

Roy, A. 2010. *Poverty Capital: Microfinance and the Making of Development*. London: Routledge.

Samuelson, P.A. 1973. *Economics*, 9th edn. New York, NY: McGraw Hill.

Sandbu, M. 2016. There is no lower bound. *Financial Times*:4 February.

Sayer, A. 1995. *Radical Political Economy: A Critique*. Oxford: Blackwell.

Schmidt, V. 2009. Putting the political back into political economy by bringing the state back in yet again. *World Politics* 61:516–46.

Seccareccia, M., and M. Lavoie. 2015. Income distribution, rentiers and their role in a capitalist economy: a Keynes–Pasinetti perspective. In *Institute for New Economic Thinking Conference, Liberté, Égalité, Fragilité*. Paris, April 2015, https://www.ineteconomics.org/uploads/papers/Seccareccia-Lavoie-Paris-INET-27-03-15.pdf.

Skidelsky, R. 2009. *Keynes: The Return of the Master*. New York, NY: Penguin.

Stein, J. 2016. A losing coalition. *Jacobin*: 14 November, https://www.jacobinmag.com/2016/11/hillary-clinton-donald-trump-working-class-election/.

Stiglitz, J.E. 2010. *Freefall: America, Free Markets, and the Sinking of the World Economy*. New York, NY: W.W. Norton.

Stiglitz, J.E. 2013a. After the financial crisis we were all Keynesians – but not for long enough. *Guardian*: 10 October, https://www.theguardian.com/business/economics-blog/2013/oct/10/financial-crisis-keynesians-eurozone-recession.

Stiglitz, J.E. 2013b. Social protection without protectionism. In *The Quest for Security: Protection Without Protectionism and the Challenge of Global Governance*, eds J.E. Stiglitz and M. Kaldor, 24–47. New York, NY: Columbia University Press.

Sum, N.-L., and B. Jessop. 2013. *Towards a Cultural Political Economy: Putting Culture in its Place in Political Economy*. Cheltenham: Edward Elgar.

Summers, L.H. 2014. The inequality puzzle. *Democracy: A Journal of Ideas* 33: http://democracyjournal.org/magazine/33/the-inequality-puzzle/.

Swagel, P. 2015. Legal, political and institutional constraints on the financial crisis policy response. *Journal of Economic Perspectives* 29 (2):107–22.

Swain, J. 2016. White, working class and angry: Ohio's working class help Trump to stunning win. *Guardian*: 9 November, https://www.theguardian.com/us-news/2016/nov/09/donald-trump-ohio-youngstown-voters.

Taylor, L. 2010. *Maynard's Revenge: The Collapse of Free Market Macroeconomics*. Cambridge, MA: Harvard University Press.

Team, T. 2016. A look at common equity tier 1 capital ratios for the largest U.S. banks.*Forbes*:6March,www.forbes.com/sites/greatspeculations/2015/03/06/a-look-at-common-equity-tier-1-capital-ratios-for-the-largest-u-s-banks/.

Temin, P., and D. Vines. 2014. *Keynes: Useful Economics for the World Economy*. Cambridge, MA: MIT Press.

The Economist. 2014. Shadow and substance. *The Economist*: 10 May, http://www.economist.com/news/special-report/21601621-banks-retreat-wake-financial-crisis-shadow-banks-are-taking-growing.

Thimann, C. 2015. The microeconomic dimensions of the eurozone crisis and why European politics cannot solve them. *Journal of Economic Perspectives* 29 (3):141–64.

Treanor, J., and Mason, R. 2017. Buy George? World's largest fund manager hires Osborne as advisor. *Guardian*: 20 January, https://www.theguardian.com/politics/2017/jan/20/george-osborne-investment-advice-blackrock-fund-manager.

Vonnegut, K. 1965. *God Bless You, Mr Rosewater*. New York, NY: Bantam.

Wainwright, T. 2009. Laying the foundations for a crisis: mapping the historico-geographical construction of residential mortgage backed securitization in the UK. *International Journal of Urban and Regional Research* 33 (2):372–88.

Wainwright, T. 2012. Number crunching: financialization and spatial strategies of risk organization. *Journal of Economic Geography* 12 (6):1267–91.

Walks, A., and B. Clifford. 2015. The political economy of mortgage securitization and the neoliberalization of housing policy in Canada. *Environment and Planning A* 47 (8):1624–42.

Wolf, M. 2008a. *Fixing Global Finance*. Baltimore: Johns Hopkins University Press.

Wolf, M. 2008b. Keynes offers us the best way to think about the financial crisis. *Financial Times*:23 December.

Wolff, R.D. 2009. Economic crisis from a socialist perspective. *Socialism and Democracy* 23 (2):3–20.

Wolff, R.D. 2010. *Capitalism Hits the Fan: The Global Economic Meltdown and What to Do About It*. Northampton, MA: Olive Branch.

Wyly, E.K., and C.S. Ponder. 2011. Gender, age, and race in subprime America. *Housing Policy Debate* 21:529–64.

Wyly, E., M. Moos, D. Hammel, and E. Kabahizi. 2009. Cartographies of race and class: mapping the class-monopoly rents of American subprime mortgage capital. *International Journal of Urban and Regional Research* 33 (2):332–54.

Zaloom, C. 2009. How to read the future: the yield curve, affect and financial prediction. *Public Culture* 21 (2):245–68.

Part I
Financial Imaginaries

2

From Time–Space Compression to Spatial Spreads

Situating Nationality in Global Financial Liquidity

Dick Bryan, Michael Rafferty and Duncan Wigan

As finance has evolved, especially from the 1980s, it has involved a transformation of economic, political and social space. Finance is liquid – it moves easily and at low cost, it can change its form with the same facility, and it is transacted in milliseconds via satellites. Money and finance have, in other words, developed new geographies (Cohen 1998, 2001; Hartmann 1998) and new imaginaries (Pryke and Allen 2000).

Predominantly this development is still being framed as an escape from pre-conceived, and invariably national, space: we have terms like 'shadow' banking (as if banking is conventionally well regulated by illuminating national states) going 'offshore' (as if capital's conventional location is 'onshore'), to 'tax havens' (places free of conventional nation-state supervision) and swirls around in 'dark pools' (traded in opaque markets). Within these issues, and the debates that surround them, there is clearly recognition of a historical transformation, yet a conventional language of space and time seems to depict change in terms of a deviation from a norm, and a loss of order. It is a struggle analytically to frame these changes in terms of their trajectories rather than their losses.

As the global range of capital becomes incorporated into analysis, we see a partial challenge to conventional conceptions of space. Space (and time too) here becomes merely of the moment; not an ordering category.

Money and Finance After the Crisis: Critical Thinking for Uncertain Times,
First Edition. Edited by Brett Christophers, Andrew Leyshon and Geoff Mann.
© 2017 John Wiley & Sons Ltd. Published 2017 by John Wiley & Sons Ltd.

The analysis of 'supply chains' and 'value chains' and even, in the case of bitcoin, 'block chains' can monitor spatially where at any point in an accumulation process a commodity, an act of production or a transaction is located, but here space, conceived as place, is a descriptor of the chain, not an ordering attribute[1] (Coe and Yeung 2015; Gereffi and Korzeniewicz 1994; Joonkoo 2010; Neilson et al. 2014[2]). Similarly, in financial markets space becomes increasingly about a framing of time, or reducing the time taken to traverse space, as competition once framed in terms of the speed of clipper ships and then aeroplanes is now framed in terms of the speed of transmissions in fibre-optic cable (Mackenzie et al. 2012). The analysis of trading in dark pools is all about manipulating the time it takes for high frequency trades to travel to market exchange mainframes (Lewis 2014).

From the 1980s David Harvey (1990), amongst others, made the bold call that globalization has annihilated space. Similarly, Gunther Teubner (1997) and others have developed a thesis that we are seeing the emergence of a 'global law without a state' (or a new 'lex mercatoria') where commercial and financial activity have 'lifted off' from national regulatory spaces. It led to a strong push-back asserting the significance of nation states and processes of accumulation within globalization (Hirst and Thompson 1998) and especially so in relation to finance (Helleiner 1996).

Perhaps, with recent innovation in finance, what we are seeing is not just a compression of time and space (Pryke and Allen 2000) but, by that compression, an opening up of new conceptions of both space and time. Indeed, the era of the algorithm and high frequency trading creates a financial reframing of all 'difference', be it spatial, temporal or any other attribute. Difference, framed financially, is a 'spread' to be priced and potentially traded. Space, defined as differentiated nationality, might then be thought of as points 'connected' financially by a spread to be traded (arbitraged).

But the fact that differences are traded does not mean they are traded away (unless we are in a hypothetical world of general equilibrium models), and differences that pertain to a dimension of national space perhaps cannot be traded away.

In this chapter, we explore how the significance of national spaces has changed as the degree and form of capital mobility has changed. The capital and finance of the 1980s is different from the capital of today, and it bears consideration that nations and nation states now matter differently from how they did in the 1950s or even the 1980s.[3] Moreover, whilst recognizing the importance of states in superintending and facilitating global finance, it is not clear that a discourse of nationality (and the escape from nationality) is the most useful way to frame global finance (Bryan 1995; Christophers 2013).

So some analytical and linguistic dilemmas emerge about the standing of nationality. Who makes up the 'national economy'? Whose debt is the 'national debt'? Which investment is 'foreign'? It is tempting to answer that there is no answer, in the light of growing mobility, liquidity and fungibility of capital. But it does not follow that nationality and the nation state are superseded categories. Indeed, they remain central organizing categories of capital.

This paper seeks to develop an analysis that attributes an ongoing financial importance to nationality. But this proposition needs to identify more than issues of sovereign state borrowing, and the tight policy constraints imposed on sovereign states by lenders.[4] It also needs to be more than a cultural and political statement of national distinctiveness which 'flows through' to finance, and it needs to say more than that in a crisis nation states are called on to underwrite 'their' banks. It needs to connect the evolving form of financial capital, and especially the rise of the derivative form, to the evolving role of nation states.

In developing this argument we begin with post-WWII balance of payments categories, not just for the sake of historical completeness, but because the attribution of nationality to capital, which was so mechanistically formalized within balance of payments accounting, has remained a pervasive influence into the twenty-first century. It is really only in the form of the derivative, as we see in the second section of the paper, that the ontological primacy of nationality in global financial analysis has been superseded by a 'spreads' conception of nationality-as-difference.

Finally, we turn to the new ways in which national governments appear to be seeking to stabilize financial markets in national terms by their capacity to regulate the role of households in financial markets, and what that might mean for an understanding of illiquid place in liquid finance

Balance of Payments and the Ontological Primacy of the Nation: A Background

Whilst balance of payments accounts have a long history, they were very limited, both in the sense of restricted data collection capacity and connection to policy variables, until the late 1930s.[5] The impact from the late 1930s of Keynesian economics was to establish in post-WWII policy an agenda that the nation state could and should manage economic performance domestically and jurisdictionally. We can readily associate this vision politically as social democratic with a conscious process of nation state control over market processes (Radice 1984).

To manage this thing called the 'national economy' via fiscal and monetary policy meant that there needed to be clear data on the level of economic activity and its components (savings, investment, consumption, etc.) and where the nation linked to the rest of the world (then simply posed as 'external'; or 'foreign' as in external balance and foreign trade). These data were compiled in the current account of the balance of payments. This account would clarify the net position on whether produced output was leaving the country, whether consumption was of local or imported goods, and whether income was leaving or entering the country.[6]

Equally, Keynesian economics required monetary policy to manage the rate of interest. Money has always been recognised as internationally mobile to some degree (Chown 1994; Cipolla 1967; Dwyer et al. 2002; French and Leyshon 2012; Leyshon and Thrift 1997), but it was critical to national economic management that there be a clearly-delineated national money system, in which the (nation-)state was central. Keynes, following the German state money school, contended that the nation state both defines what is accepted as money and produces the national money supply. It thereby controls the quantity of money in circulation and can adjust that quantity to impact the rate of interest. With money able to move in and out the country (at first in restricted ways), the state needed to know whether there was a net entry or exit, so that local supply could be adjusted to secure the desired quantity in national circulation so as to manage the domestic rate of interest. This information could be gleaned from the capital account of the balance of payments.

Critically, this balance of payments accounting is predicated on the notion that the national and the international could (and should) be clearly differentiated. Conceptually, it presumes the nation-as-economy as ontologically prior to its analytical agenda. For Keynes, this differentiation was to be enforced by policy: by trade controls for the current account, fixed exchange rates and controls on international credit and investment in the capital account. By means of these policies the nationality of economic activity was consciously delineated and a national economic community imagined.

These accounts ruled the interpretation of international trade and financial flows in the post-war period. They informed national governments of policy requirements to ensure 'national balance' in relation to the 'rest of the world' and they were (and still are) used for policy formulation by international organizations like the IMF, the World Bank and the WTO, though, as we will see, with increasing difficulty.

Things changed from the 1970s, with the end of the global 'long boom' and the termination of the post-war Bretton Woods Agreement on international finance. An oil crisis, unemployment and inflation throughout the advanced capitalist economies became a turning point materially and in economic policy; indeed it is the benchmark change from which it is said 'neoliberalism' emerged (Helleiner 1996).

Financially, this turning point was about the rise of Eurofinance markets: the trading of currencies outside their country of issue. Interest rates and exchange rates that differed significantly from domestic rates created ready gains from interest rate and exchange rate arbitrage. Yet for financial institutions from countries subject to the Bretton Woods Agreement, trade in these Eurofinance markets was officially prohibited. The aspiration of bankers to push down the wall constraining cross-national movements of capital grew, and indeed what became known as 'deregulation' emerged from the early 1970s as a direct and indirect result of that aspiration (Burn 2006; Christophers 2013; Germain 1997; Helleiner 1996).

Yet so entrenched had become the economic categories of national income accounts and the balance of payments that the economic taxonomy remained convention in both economic theory and policy, even though the policies which performed the taxonomy waned. The spatial economic processes and the categories to measure them started to become incoherent. A remarkable conceptual void opened up at this critical turning point, one that continues to challenge economic analysis more and more acutely as capital becomes more and more liquid with the rise of financial derivative markets.

In relation to the current account of the balance of payments, the adherence to a national taxonomy of economic activity initially manifested in significant but partially manageable ways. As global supply chains of commodity production became more complex, it was difficult to say just where any item was produced: how much local content must a good have before it can be classified as 'local'? But the question did not present directly as a policy question (Coe et al. 2007; Desai 2008; Gereffi et al. 2005). If the raw materials of a good are local, but it is manufactured internationally and then re-imported, trade data can still monitor (transfer pricing issues aside) the net position on local value creation.[7] Within the current account, the more complex question was income flows, which themselves predominantly attach to previous financial transactions (changes in ownership, debt, etc.) recorded in the capital account, although they can also be, for example remittances of wages.

It was in the capital account of the balance of payments, recording financial transactions, that decisive challenges to a national taxonomy of capital manifested. It showed up in the taxonomy of mobile capital in the forms of multinational corporations and money.

Nationality and International Investment: Multinational Corporations

In the 1980s there was a sustained debate between Laura Tyson and Robert Reich, both of whom later became key economic figures in the Clinton administration: Tyson as head of the president's council of economic advisors and Reich as Secretary of Labor. The debate, predominantly in the *Harvard Business Review* and *The American Prospect*, was called 'Who is US?' (Reich 1990, 1991b, 2005; Tyson 1991). The title provides a neat play on whether US is a pronoun or a noun (us as community or the United States as sovereign nation). The issue of debate was the extent of internationally mobile investment by multinational corporations: how much US investment there is offshore, and how much 'foreign' investment there is in the US. So, Tyson and Reich debated, should the US-as-nation be economically defined less in terms of a geographical space (the conventionally-understood nation), or reframe nationality either in terms of nationality of asset ownership, or behaving nationally (living, investing and producing with the US)? An ownership redefinition favoured by Tyson would include US-owned corporate assets abroad as constituent of the United States economy, and exclude 'foreign-owned' corporate assets located in the US. A behavioural framing, favoured by Reich, would pay attention to the corporations based on where they locate key investments and activities (R&D, headquarter functions, etc.), rather than the nationality of corporate ownership. To add extra spice to the conceptual dilemma, in the mid-1980s, when the 'geographical US' was running a growing trade deficit, and the talk was that the US was losing its global competitiveness, one calculation showed that the 'ownership US' was running surpluses (Julius 1990).

It seemed that these different ways of defining economic nationality were pointing in different directions. The US as a geographical space may have been losing competitiveness as a site of production, but US-owned capital was more competitive than ever: it was just expressing that competitiveness by being spatially mobile.[8] For balance of payments analysis, however, this challenged the basic taxonomy:

> With integrated capital markets and floating exchange rates the balance of payments has lost its former role as the authoritative summary of a nation's external position. It is merely one way of sorting economic transactions into categories that present a balance sheet of residence-based, currency-specific trade and capital flows. (Julius 1990: 85)[9]

There followed a series of formal inquiries into the problems of balance of payments accounting taxonomy.[10] The consequence of these reports was to codify the ramifications of these different definitions of nationality[11] and the result was that from 1995 the US Bureau of Economic Analysis in the Department of Commerce started publishing two sets of balance of payments data – one (conventionally) geographically defined, another ownership defined (Mataloni 1995; 1998).[12] But which was the 'real' US: who is us, and which us defines economic nationality?

That debate could have no resolution, for it rested on definitional issues, albeit that it is interesting to recognize that as the denationalization of investment has extended over the subsequent thirty years, and the divide between geographical nation and investment nation has widened, the debate itself has almost disappeared.[13]

That debate may have dimmed, but it does have new iterations, especially in relation to multinational financial institutions, where offshore branch structures grew from the 1970s, and where the national location of financial 'production' became increasingly ambiguous. Not only do the balance of payments categories of debt, direct and portfolio investment become too limited to classify the novel sorts of assets moved by investment banks and hedge funds, but processes now called 'shadow banking' and the use of high-frequency trading outside of formal exchanges (so-called dark pools (Lewis 2014)), make it difficult to know how to assign a spatial dimension to financial transactions. Indeed, shadow banking may be thought of as creating not just spatial ambiguity of national corporate classification, but also category ambiguity in the classification of different asset types. Both feature significant taxation implications: the former for where (and whether) taxation will be paid (Sharman 2010); the latter for which particular forms (and rates) of tax will be paid (Bryan, Rafferty and Wigan 2016). We explore this latter issue shortly.

When these developments are framed in the discourse of the ontological primacy of nations, the focus is on the apparent avoidance of the consequences of national regulations by the use of tax havens, and so the loss of legitimate revenues to national treasuries (although it should be noted that tax havens can only have status because they too have the national sovereignty to offer tax and other concessions). Nonetheless, the ongoing privileging of the discourse of nationality is leading to high-profile international forums on international tax reform (e.g., G20 2009, OECD 2013) and to calls for a financial transactions tax, to slow the mobility of capital and wrest control back to national systems.[14]

The enthusiastic advocacy of transactions taxes by political leaders and NGOs, especially in the immediate aftermath of the global financial crisis, but the absence of subsequent implementation, can be read as an ideological reassertion of state sovereignty in a time of crisis. This is an issue to which we will return later in the chapter.

National Money and International Finance

In the period after the slump of the early 1970s, production and trade dominated policy formation and analysis. With the deindustrialization of critical parts of the advanced capitalist countries, focus turned to rebuilding those economies on the principles of flexible, high productivity labour. National competitiveness, framed in the managerialist language of 'competitive advantage' (Porter 1990) saw advanced capitalist countries trying to attract mobile capital back to their shores by the promise of state-nurturing (grants, tax breaks) of high technology industries and skilled labour forces. In this framework, there was the recognition that capital (especially multinational corporations, but credit too) had become global, but the policy agenda was one of national capture (or in geographic terms spatial refixing).

Whilst this 'productionist' agenda was (and remains) ongoing, by the 1980s attention gradually turned to finance as the critical frontier in the global mobility of capital. How did the contradiction between national taxonomy and international mobility play out in relation to finance? Following Keynes, post-war economic convention depicts a nationally delineated money system (and often a national money supply), issued by the state and regulated by the monetary policy of the national central bank (Europe became a telling exception here, but for the moment we can think of the European monetary union as a single entity).

In the midst of the 1970s crisis of Keynesianism, and the threat of 'stagflation', monetarism emerged as the new policy wisdom. It centred on the proposition that national economic policy should be passive, not counter-cyclical like Keynes advocated, and focused on steady control of the supply of national money as the means to support non-inflationary growth. But despite its refutation of Keynesian policy, it retained from Keynesianism the notion that there is a clearly delineated national money. But while for Keynes this was conceived in the context of a 'closed' economy, Friedman believed the state could 'sterilize' the national money supply (that is, neutralize by making additions and subtractions to local money supply in the light of net international inflows or outflows). Financial innovation, as we will see shortly, made this impossible to implement effectively.

Accordingly, many countries that adopted formal monetarist theory did so only briefly, for it became readily apparent that the monetary aggregates targeted for policy were not stable once they became the objects of policy targeting (Goodhart 1981: 116[15]). It became clear that there was no single, clear delineation of a 'national' money supply, and one of the reasons for that was that, especially with the rise of Eurocurrency markets, the spatial circulation of money and the national jurisdiction of its regulation became de-linked. Today, central banks now understand that 'private' banks control the issue of money through their lending policies (e.g., McLeay et al. 2014), and national central banks can influence only interest rates, and even then it is but a partial influence because of the global mobility of credit.[16]

From the 1990s onwards, at roughly the same time that 'national competitiveness' in industry policy gave way to more individualized, market agendas of trade, productivity and profitability, finance too transformed its attachment to nationality. The question had to be asked (at least in retrospect) whether there is a 'national' money system when the movement of liquid funds across borders has no real costs: where assets can be held in foreign currency within a nation, and 'foreigners' want access to the nation's currency for the purposes of portfolio diversification. Is there still significance to a category of 'national' savings when local companies borrow internationally and 'foreign' companies, even those with no local investments, borrow in the local market, or at least in local currency? What does national debt mean when it is increasingly a reflection of private (corporate and bank) borrowing decisions about how to fund investment, and thereby not a statement about 'national' borrowing per se? In equity investment, legal facilitation and changes in communications technology meant that companies of many sizes were starting to invest 'offshore': to be a multinational corporation means no longer being exceptional (a giant conglomerate with a key technology, market position or brand with the capacity to overcome the costs of international commerce), but simply to be networked and spatially linked beyond a single country. And, with alliances, outsourcing, licensing, franchising and so on, control of operations abroad became possible with little or no outflow of capital from the home country. Furthermore, in portfolio investment, with the rise of pension funds, we saw purchases beyond the 'local' stockmarket, in the name of risk diversification. So (once) national bond markets, stock markets and loans markets started to lose national particularity – never entirely, for there is always a residual local 'bias' (as it is called) for reasons of both familiarity and exchange rate risk, but sufficiently to challenge an image of nationality tied to autochthonous accumulation.

Derivatives: Beyond a National Taxonomy

The developments depicted above occurred not just because of a shift in the regulatory regime in many countries that enabled rapid increases in cross-border movement of goods and services, income, credit and equity, but also because finance was itself changing, both quantitatively and in its form and meaning. We refer here specifically to the rise of financial derivatives and related leveraged products like collateralized debt obligations (CDOs). To place this development in historical context, we return to the historical turning point at the end of the 1970s that proved pivotal for both trade and investment flows. The retreat of the state from active national economic management we have described earlier can be framed differently, in a way that points to finance.

The shift away from Keynesian policies meant that the state withdrew from managing the connections between the present and the future (Negri 1967). It would be finance, through insurance and futures (and options) markets, which would now have to take responsibility for linking the present to the uncertain future. In these markets the links to the future are, however, performed through a myriad of transactions as largely individual, not collective responsibilities. This observation, in itself, signals a change in the nature of economic nationality away from collective identity. We will return to this issue later.

For many, there was an automatic jump to the proposition that if these markets had transcended national bounds, that they had therefore also lost a national attachment. It tied directly to the idea of derivatives as 'mere' speculation, having no attachment to 'real' capital accumulation.

It was not the case at all. Exchange rate and interest rate derivatives, and swaps markets in particular (Leyshon 1996) are generally conceived in arbitraging national spaces of money: US against Japanese currencies or three month peso or pound interest rates. Differences in exchange rates or interest rates between currencies represent what is conceived in markets as a spread (will the gap between two national currencies or interest rates widen or narrow over the next three months? etc.). Moreover, derivatives on interest rates and exchange rates make up 92 per cent of the over the counter, and 99 per cent of the exchange traded global derivative trade (BIS 2015; BIS 2015a). But here we see nationality shift from a policy-coherent place where production of goods and services locates to a tradable attribute of an asset.

The other critical development of the early 1970s was the publication of the Black–Scholes–Merton options pricing model and the invention of hand-held computers which, in combination, enabled a real-time so-called 'scientific' trading of risk. Retrospectively, there is debate as to whether that model 'works': whether it was performative and then counter performative (MacKenzie 2006), but the critical point is the market's initial confidence in the formula as a precise device of liquid risk calculation. Its specific formulation calculated the price of a hedge to preserve the value of an underlying asset portfolio. It meant some critical financial volatilities once hedged by the nation state through regulation could now be hedged privately through the market.

The capacity to trade risks of the future was therefore rapidly becoming an integral feature of financial markets, and this saw the growth of derivative markets, often as a strategic response to national capital controls. Mehrling (2011) contends that the development of parallel loans, the precursor to swaps, was stimulated by US capital controls and the tax wedge imposed as part of those controls.

The meaning and significance of the rise of derivative markets does not need extensive consideration in the current context. One basic point, however, warrants emphasis. Derivative markets trade exposures to the performance of underlying assets, but without (necessarily) trading the underlying assets themselves. This separation is integral to their role in hedging and speculating on underlying assets: by trading exposure to the change in price of an asset (or movement in an index) the trader holds a leveraged position on the value of an underlying asset. Derivative trading does not require physical ownership of the commodity or asset, and the exposure and the thing are de-linked. That de-linking is not just physical and spatial, but also national and jurisdictional in its nature.

Especially with the development of algorithms for high frequency trading, the objective of derivative activity can be respecified as the trading of volatility. If Black–Scholes was initially framed as an agenda for hedging (balance and preservation), high frequency trading is about arbitrage: looking for opportunities, which may only last hundredths of a second, to trade on spatial and temporal 'mispricings'. Indeed, Black–Scholes is now thought of primarily not as a measure of the price of a hedge but as the measure of market volatility (Derman and Miller 2016).[17]

A fragility immediately appears, and it is not simply about the dangers of 'speculation'. It is that derivative markets trade volatility, yet they need volatility to stay within limits defined by norms of market stress testing. When volatility is too extreme (beyond those calculated in the models), the algorithms fail (MacKenzie 2011). This has been a not-infrequent phenomenon of liquid financial markets, expressed in the 1998 crash of Scholes and Merton's hedge fund Long Term Capital Management and in the toxic mortgage-backed securities and CDOs, and then credit default swaps of the global financial crisis. Depicted by Taleb (2007) as 'fat tails' (on probability distributions), these unexpectedly volatile events suggest a need for moderation (regulation) in the operation of financial markets.

So a dilemma follows the fragility: instabilities which are innately international can only be formally regulated nationally, or by cooperation between national states, because there are no formally-constituted global authorities (Helleiner and Pagliari 2011; Tsingou 2015; Grinberg 2015[18]). Moreover, many of these national regulations are implemented by concerted nation-state interventions in financial markets where those states have regulatory capacity only as major, asset-rich traders in the market rather than because of a sovereign power of decree. This represents a very different conception of nation-state and of regulation from that which applied in the era of post-WWII capital controls.

Balance of Payments after Derivatives

Largely in response to high-volume derivative transactions, the official procedures for Balance of Payments accounting were changed by the IMF (1993) and OECD (1996). The change saw the 'capital account' split into a 'financial account' and a capital account', and derivatives were to be recorded as notional amounts outstanding at the end of an accounting period (IMF 1997). This was some recognition of the impact derivatives, but a snapshot of a derivative position at a point in time (31 December) provided as static conception of highly liquid capital.

Developments in derivative markets proved an impossible challenge to balance of payments accounting, and hence for the economic representation of the nation vis-à-vis global derivative markets. Four challenges stand out. First, derivative trades are frequently cross-national financial transactions, but many thought they have no impact on national money supply, and hence national policy. Hence there was no policy challenge, for derivative positions trading different perspectives on risk, will sum to zero. This would prove a contentious view.

Second, derivative markets involve rapid turnover and high volume trading. Current high-frequency trading involves buying and selling over fractions-of-a-second intervals. The capital account of the balance of payments, on the other hand, measures only changes in the value of stocks over an accounting period, with a year the standard measure. Measuring stocks at a point in time (say, midnight on 31 December) to be compared with the same time in twelve months has meaning for measuring changes in stocks of direct and portfolio investment, but not for the real time fluidity of derivatives and other forms of finance. One second later, the effect of high frequency trading in derivatives may mean that a stock balance may be significantly different.

The third challenge to balance of payments accounting is possibly the most significant, at least with respect to an economic conception of spatial nationality. The possibilities provided by trading exposures to asset performance extend beyond hedging an underlying asset. Balance of payments seek meaning in underlying assets – investments, savings, consumption, etc. Derivatives open up the possibility of a disjuncture between what appears to be the case in terms of underlying assets and where financial exposures to these underlying assets are located. This was shown starkly in the case of Enron, where its derivative positions were undertaken by a network of subsidiaries (so-called special purpose entities or vehicles, with their office registered in multiple-off shore jurisdictions) and too complex to be disentangled by auditors. The same applies, indeed on a broader (though not intentionally deceptive) scale, for countries.

This is especially the case in relation to the issue of the nationality of equity and debt. Equities involve (at least) three attributes: exposures to capital gains/losses, to dividends and share voting rights. In derivative markets these can each be traded separately. Products like total return swaps mean that where a share is formally 'owned' (voting rights) can be separated from where dividends accrue, and from where capital gains/losses are carried. Henry Hu, inaugural Director of the Division of Risk, Strategy and Financial Innovation at the US Securities and Exchange Commission calls it 'equity decoupling' (Hu and Black 2007). Decoupling means that nothing about the 'operation' of a corporation can be deduced from where it is formally owned. So a company may formally be 'locally' owned for national policy (e.g., subsidy or foreign investment control) purposes, but all exposures to the performance of ownership are held offshore.

Similarly, distinct analytical categories like debt and equity have to be seen as historically specific and potentially outmoded,[19] with products like preference shares having attributes of both equity (no expiry date) and debt (predetermined yield and no voting rights). Finance may enter a country as equity (foreign investment) but the exposure swapped into local debt. National policies that may be implemented to encourage long-term capital inflow and discourage short-term capital flows may be side-stepped by the owner of the long-term investment swapping their exposure for short-term, speculative positions. All this is entirely missed by post-war balance of payments categories that could not have imagined the possibilities and consequences of financial derivatives.

The fourth challenge to balance of payments categories is technically small, but of great significance. Financial markets developed products that broke down the distinction between debt and equity. This includes products like preference shares and total return swaps that have attributes of debt, such as a predetermined rate of return, but, like equities, no expiry date. Combined with equity decoupling, the impact of these sorts of products is to confuse the notion of equity ownership, and hence corporate ownership generally. The Tyson/Reich debates of the 1980s, predicated on a clear delineation of nationality of ownership, is no longer possible.

Indeed, with financial transactions being so large in volume, strategic risk management positions of large corporations and financial institutions can come to dominate balance of payments data and so give a completely false sense of national economic processes. IMF economist Peter Garber has succinctly summarized the problem. He contends that financial derivatives provide firms with an:

> increased ability to separate and market risks ... Coupled with the existence of weak financial systems and the inherent opaqueness of derivative positions due to obsolete accounting systems ... derivatives can be used to leverage financial safety nets ... Often, such activity must move offshore to evade detection and naturally generates a gross international capital flow. Moreover, derivatives can be used readily to evade onshore prudential regulation and capital and exchange control, thereby generating yet more measured capital flows ... (Garber 1998: 2)

Nation and Place in Liquid Finance

In the light of the financial and industrial developments described above, it would be hard to claim that the nation remains a clear economic category. Our analysis has, to this point, sought to frame economists' visions of the nation as at best anachronistic and more strongly as misleading. Economists mistook a historically-specific and policy-dependent construction of economic nationality as if it were an ahistorical category.[20] It was a framing that was a grand vision of post-war state economic management and social democracy that has, in many ways, run its course in the light of more recent developments in capital accumulation. Terms like 'neoliberalism', 'globalization' and 'financialization', while no doubt easy tags with imprecise meanings, nonetheless point sufficiently for our purposes to the sorts of momentums that shattered the post-war basis of that economic nationality.

Yet in the major international economic institutions like the IMF and Bank for International Settlements, there is still a real reluctance to let go of balance of payments data. IMF country analyses still constitute the 'external sector' as a measure of performance of national countries, and it features centrally in their national risk assessments. This is, no doubt, part of what sees the IMF being widely depicted as the epitome of a 'neoliberal' institution.

Yet in the Bank for International Settlements, there are the beginnings of doubt. Claudio Borio, long-time initiator of new economic thinking in the Bank has, for several years, been challenging the use of balance of payments data and the 'residency principle' that underlies these data. For Borio (2015), this is a matter of the need to 'take financing more seriously'. It rests on the recognition that a savings constraint on a nation is not the same as a financing constraint. The former is a national stock; the latter is a global flow, and in the current era of finance, it is the latter that counts; not the former. Nonetheless, in Europe, and most conspicuously in the EU negotiations with the Tsipras Government of Greece during 2015, it is the lack of savings that has been used to rationalize policies of austerity in the expectation that it will see Greek savings converted to a financial flow for bank repayment. In effect, it is apparent that balance of payments ways of thinking, whilst financially anachronistic, remain potent tools of international political control.

But nor should the impression be created that nationality of finance is merely a neoliberal device (although nation-state bail-outs of banks in the midst of the global financial crisis may look like further evidence of that claim). So in what ways do national spaces, and the social and political ties engendered therein, matter financially? Two dimensions warrant emphasis: financial market need for regulation and the associated opportunities for regulatory arbitrage, and nation-state demands of financial citizenship. Each warrants brief consideration.

Regulatory arbitrage

Highly liquid and global finance, which might at first appear to annihilate space (Harvey 1990) in its circulation and have 'lifted off' from national regulatory spaces (Teubner 1997), actually needs to consistently stop moving in order to execute trades and declare profits. Robert Wai (2002, 2008) and John Biggins (2012) contend that 'lift off' can never be entirely achieved in international transactions. Institutions and transactions have to 'touch down' in some jurisdictions to achieve certain benefits (such as property rights protection, etc.). Rather, in a plural and fragmented jurisdictional world, where each jurisdiction is 'rife with contradictions, gaps and ambiguities', some partial lift-off can be achieved. But financial institutions and corporations are still faced with (now) strategic choices about which transaction types 'touch-down' where, when and in what institutional form. Biggins calls this forum or jurisdictional platform selection decision-making 'targeted touchdown'.

Partial lift-off and targeted touch-down are of course metaphors,[21] but they point to a disjuncture between the times and spaces being produced by modern finance. They signal that space and place are important, but as strategic variables, not as ontologically prior categories. The central insight here is that the financially transforming process of 'unbundling' corporations into a range of attributes or activities takes this disjuncture as the foundation of arbitrage strategies. In the process, we see that the disjunctures and ambiguities with respect to conventional categories acquire an alternative coherence in the discourse of financial calculation. Discontinuous jurisdictional space and temporal ambiguity are spreads to be strategically leveraged and traded within the flow of financial products.

National citizenship

Robert Reich, of the original 'Who is Us' debate, contended that the essence of economic nationality in a globally fluid world is labour – the one essentially non-internationally-mobile factor of production. Reich's call (1991: 301) is for America (and by implication other nations) to be defined in terms of human productive capacity, and for that capacity to be nurtured as a collective (national) rather than private (corporate) agenda.

In the world of twenty-first century finance, his insight can be reframed: that it is the immobility of people compared with capital that gives nationality meaning. However, we need to shift beyond the coupling of 'labour' and 'production' to something broader. Reinterpreting Reich's emphasis on immobility invokes Nobel-Prize-winning economist Robert Shiller's (2003: 9) insight that:

> Far more important to the world's economies than the stock markets are wage and salary incomes and other non-financial sources of livelihood such as the economic value of our homes and apartments. That is where the bulk of our wealth is found.

Critically, this housing (and household) wealth is innately place-specific. It is illiquid. But one of the frontiers of post-1970s financial innovation has been to build liquid financial assets on the platform of illiquid housing and other household payments. The most conspicuous of these are mortgage-backed securities (MBS) but other household payments, particularly rent, debt repayments (credit card and student loan) and utilities payments (phone, gas, electricity, water) are all being securitized, and sold as assets in global financial markets. These are called asset-backed securities and, in a different form, collateralized debt obligations (CDOs).

This process of securitization involves bundling up household contractual payment obligations, transferring the ownership of the income stream to Special Purpose Vehicles (SPVs) for ultimate sale into global financial markets. Wyly et al. (2009), following Harvey, refer to this securitization as the globalization of class-monopoly rents. It is not new – it was already conspicuous in relation to mortgages in the 1980s and 1990s (Pryke and Lee 1995; Leyshon 1996) – but it came to prominence in the global financial crisis, for the well-known trigger for the crisis was the unexpected (albeit in retrospect unsurprising) high level of defaults on subprime mortgage-backed securities.

In the aftermath of the crisis, it is notable that regulatory change to prevent repeats of such crashes has focused not so much on the assertion of dramatic new controls over traditional banks and their use of SPVs. Such controls have been at best modest and these shadow banking practices are largely beyond the regulatory capacity of any single nation state. Rather, the focus has been on the financial capacities of households, which are innately tied to place. Accordingly, we have seen a regulatory agenda that is predicated on the immobility of households: we have seen changes to personal bankruptcy laws in the US, and an emerging surveillance of the spending practices and financial viability of households, so as to manage default risk and its impact on securitized household payments. More broadly, we see the rise of a social culture of 'responsibilization', that individuals and households are expected to manage a range of social and life course risks (through financial markets) that were formerly managed by states and employers (Martin 2002; Shamir 2008; Beggs et al. 2014)

So instead of the state playing the 'Keynesian' role of stabilizer and insurer, we see the state demanding that individuals take on these precise roles,[22] as a modern iteration of what E.P. Thompson called a 'moral economy',[23] with the state itself in the role of 'managing expectations'. We see a morality of staying 'on contract', and 'on payment', even at personal cost: an expectation that applies only to individuals as national citizens, but cannot be made of corporations which are not conceived in the domain of morality. It is illegal for corporations to trade while insolvent, but households with negative net assets (debt greater than total asset value) are expected legally and culturally to struggle to meet their payments.

Here we can start to see the reconstruction of a nationality in finance, not in terms of the axiomatic innate nationality of capital or financial institutions, but in the development of a financialized citizenry, which anchors the nationally denominated financial claims circulating globally.

Conclusion

As we track the economic and social basis of nationality and place in relation to the evolution of the techniques of finance, a transformation of both is apparent. The post-war period of strict capital controls and tightly regulated banking was tied to (broadly) social democratic goals of nation building and social balance, conceived in state guarantees of jobs, wages, welfare and managed inequality. The dramatic social and economic changes of the 1970s and (especially in relation to finance)

1980s generated a breakdown in that collectivist agenda, and the rise of globally-integrated, competitive agendas, often termed 'neoliberalism'.

That much is well known. But there is little consideration of the way in which changes in the techniques and recording of financial flows impacted on this transformation. Too often, we hear that finance was 'deregulated' and states simply ceded control to 'markets'.

Yet, ironically, finance today is probably the most state-subsidized business in the advanced capitalist countries – from a combination of bail-outs and underwriting to ensure confidence in the sector, to the fees that come from people being increasingly forced to borrow in order to get a home or education, and to save for old age. In the US and Europe the monetary policy of quantitative easing sees the state guarantee purchases of financial products. This is not a process of 'deregulation', but a rebuilding of the national meanings of globally-mobile finance.

What has changed critically in this process is the way in which citizens are being called on to underwrite the stability of finance. The IMF (2005: 5) has described households as the financial system's 'shock absorber of last resort', and it plays this role because, unlike finance, households are illiquid. They are place-specific and time-specific too, in the sense that they cannot readily defer all economic activity or future rounds of contract payments. Because of this illiquidity they can be closely regulated by nation states so that they continue to systematically absorb financial risk. This is a process that is, and can only be, innately national.

Notes

1 There is an interesting parallel here with Arrow–Debreu general equilibrium, where a commodity in each place and each time period must be constituted as a discrete commodity class, so that each 'commodity' can have a discrete equilibrium price. Thanks to Geoff Mann for making this parallel.
2 Seabrooke and Wigan (2014) suggest a range of pay-offs in incorporating finance, law and accounting into the concerns of GVC scholars, including the spatial implications of doing so.
3 There is a 'varieties of capitalism' literature that focuses on the relative roles of states and markets in economic coordination. Our issue here, whilst not exclusive of that framing, addresses the take-up and impact of financial innovation.
4 Particularly notable here are the IMF lending conditions to states in the 1997 Asian Financial Crisis and the constraints on sovereign state borrowing, and especially conditions recently negotiated with Greece, in the current Euro-crisis.
5 Patinkin (1976) for instance contended that pre-General Theory national accounts were constructed in a way consistent with the needs of pre-General Theory business cycle theory for statistical verification.

6   Note that income flows, including interest payments and dividends linked to finance, are shown in the current account because they are constituted as current income a flow rather than a stock figure, even though this was often a statistical artefact – as in repatriated and reinvested earnings of foreign subsidiaries.

7   Whilst the accounting procedure may be in a technical sense accurate, as trade grows in proportion to local production, so too does the meaning of the national economy.

8   Julius recalculated the US trade balance for the year 1986. The standard accounts, based on the national residence of production, showed a deficit of $US 144.4 billion. When data are recalculated on the basis of national ownership of production, US companies' global operations show a trade surplus of $US 56.7 billion (1990: 81). A comparable calculation for the 1987 trade balance in Kester (1992: 40) shows a balance of payments based deficit of $US 148 billion, compared with an ownership based surplus of $US 64 billion. In effect, US capital was expressing its competitive success by expanding its production globally. Lipsey and Kravis (1985, 1987) estimated that the share of world exports accounted for by US multinationals had remained fairly constant from the mid-1960s.

9   For essentially the same reasons, Lipsey and Kravis concluded that 'a distinction must be made between the competitiveness of the United States as a location and the competitiveness of US MEs [multinational enterprises] which export from both the United States and from foreign locations' (1992: 194).

10   The most notable led to reports of the US National Research Council Committee on National Statistics chaired by development economist Robert Baldwin (Kester, 1992, 1995) The panel involved all Federal agencies concerned with trade and statistics collection (Kester 1992).

11   The panel's first recommendation (Kester 1992: 5) was for a 'supplemental statistical framework that integrates balance-of-payments data and data on affiliates' operations at home and abroad ... to better reflect the link between trade and foreign direct investment'.

12   Those 'supplemental' data are now published annually in the Survey of Current Business, as national accounting-style data on the global operation of US corporations.

13   Almost disappeared, because in 2005, Reich and Tyson reprised their 1992 debate. Tyson said that her argument for nationality of ownership remained valid. She cited statistics that the 'at-home' shares of employment and capital spending of US multinationals had remained quite stable. Tyson concluded, 'For the foreseeable future, the ability of the American economy to create good jobs for American workers will continue to depend on the competitiveness of American companies' (2005: 2). She did make one concession to Reich, which was that the overarching goal of national policy was now to provide a competitive place for global companies – domestic and 'foreign'. Reich, on the other hand, contended that his 1992 prediction of American-based firms becoming more important than US owned firms was if anything understated. He cited the example of Toyota which had

become the second largest car manufacturer in the US, and the statistic that almost half of US imports now came from US companies' operations abroad. Moreover, Reich concluded that he had not fully comprehended the political dimension of his declining economic nationality argument. He failed to predict that '(b)ig American-based corporations, becoming less dependent on the productivity of Americans, would use their muscle to reduce taxes, thereby preventing the needed public investment' (2005: 2).

14 This sort of tax was being advocated by a number of national leaders – for example, the French President Nicolas Sarkozy and the British Prime Minister Gordon Brown – in the immediate aftermath of the Global Financial Crisis.

15 Goodhart's law, as it came to be known, was first presented in 1975. It states, simply, that when a measure becomes a target, it ceases to be a good measure.

16 Moreover, in the 2010s, with interest rates close to or at zero, central banks cannot use interest rate policy to generate growth, and have reverted to asset buying (quantitative easing) to add cash to the economy.

17 After 1987, Black–Scholes prices were not 'working' to hedge the value of a portfolio. Mackenzie frames this as counter-performativity. In the markets, it was interpreted as a statement that people were mis-pricing market volatility (the only variable in the Black–Scholes model which must be calculated, rather than simply compiled from available data). So if Black–Scholes prices were 'wrong', it could only be because volatility was miscalculated. Hence in not 'working' to price a hedge, the model became a means to retrospectively calculate volatility, or 'implied volatility'.

18 Grinberg (2015) makes the useful distinction between agenda setters (the G20), standard setters (such as the Basle Committee) and monitors and enablers (such as the IMF and World Bank). But none acts with the authority to enforce.

19 Myron Scholes (1997: 146–7), winner of the 1997 Nobel Prize for Economic Science for the Black/Scholes/Merton option pricing model said prophetically in his acceptance speech:

Standard debt and equity contracts are institutional arrangements or boxes. They provide particular cash flows to investors with their own particular risk and return characteristics. These institutional arrangements survive only because they provide lower cost solutions than competing alternative arrangements … Time will continue to blur the distinctions between debt and equity.

20 Mitchell (2002) argues persuasively that the origins of this stems from imperial economic governance.

21 We are not short of spatial concepts and metaphors in this domain, from spaces of flows, fixity and motion, plateaus, action at a distance, and space-time compression (Castells 1996; Brenner 1998; Harvey 1990; Latour 1987; Deleuze and Guattari 1987). Sheppard (2002) provides a useful survey.

22   Indeed, it warrants noting in relation to money that the post-financial-crisis state policy in the United States of Quantitative Easing now sees the Federal Reserve purchasing and retaining large quantities of mortgage-backed securities – indeed more of these than of Treasury bonds.

23   The term is anthropological, relating to the compliant behaviour of peasants on the edge of starvation (e.g., Scott 1976), but now more popularly attributed to Thompson (1991). Clark and Newman (2012) apply the concept to the current period.

References

Beggs, M., D. Bryan,and M. Rafferty. 2014. Shoplifters of the world unite! Law and culture in financialized times. *Cultural Studies* 28 (5–6): 976–96.

Biggins, J. 2012. 'Targeted touchdown' and 'partial liftoff': Post crisis dispute resolution in the OTC derivatives markets and the challenge for ISDA. *German Law Journal* 13 (12): 1299–328.

BIS 2015. Statistical release: OTC derivatives statistics at end June 2015. Basle: Bank for International Settlements.

BIS 2015a. Exchange trade derivatives statistics. Basle: Bank for International Settlements.

Borio, C., and P. Disyatat. 2015. Capital flows and the current account: Taking finance (more) seriously. BIS Working Papers No. 525, October.

Brenner, N. 1998. Between fixity and motion: accumulation, territorial organization, and the historical geography of spatial scales. *Environment and Planning D: Society and Space* 16 (4): 459–81.

Bryan, D. 1995. *Chase Across the Globe: International Accumulation and the Contradictions for Nation States.* Boulder, CO: Westview Press.

Bryan, D., M. Rafferty,and D. Wigan. 2016. Politics, time and space in the era of shadow banking. *Review of International Political Economy* 23 (6):941–66.

Burn, G. 2006. *The Re-emergence of Global Finance.* Basingstoke; New York: Palgrave Macmillan.

Castells, M. 1996. *The Rise of the Network Society.* Oxford: Blackwell.

Cipolla, C.M. 1967. *Money, Prices, and Civilization in the Mediterranean World, Fifth to Seventeenth Century.* New York: Gordian Press.

Chown, J.F. 1994. *The History of Money from AD 800.* London: Routledge.

Christophers, B. 2013. *Banking Across Boundaries: Placing Finance in Capitalism.* Malden, MA: Wiley-Blackwell.

Clarke, J., and J Newman. 2012. The alchemy of austerity. In *Critical Social Policy* 32 (3):299–319.

Coe, N., P. Kelly and H.W.C. Yeung. 2007. *Economic Geography: A Contemporary Introduction.* Oxford: Blackwell.

Coe, N. and H.W.C. Yeung. 2015. *Global Production Networks: Theorizing Economic Development in an Interconnected World.* Oxford: Oxford University Press.

Cohen, B. 1998. *The Geography of Money.* Ithaca, NY: Cornell University Press.

Cohen, B. 2001. *The New Geography of Money*. Princeton: Princeton University Press.

Deleuze, G. and F. Guattari. 1987. *A Thousand Plateaus: Capitalism and Schizophrenia*, Minneapolis: Minnesota University Press.

Derman, E., and M. Miller. 2016. *The Volatility Smile*. Hoboken,NJ: Wiley.

Desai, M. 2008. The decentering of the global firm. Working Paper 09-054, Harvard Business School.

Dwyer, G. and J. Lothian. 2002. International money and common currencies in historical perspective. Federal Reserve Bank of Atlanta Working Paper No. 2002–7, Atlanta, Georgia.

French S. and A. Leyshon. 2012. Dead pledges: mortgaging time and space. In K. Knorr-Cetina, and A. Preda (eds). *The Oxford Handbook of the Sociology of Finance*. Oxford: Oxford University Press.

G20.2009. Global plan for recovery and reform. Statement issued by the G20 leaders, 2 April, London, available from: http://www.cfr.org/financial-crises/g20-global-plan-recovery-reformapril-2009/p19017 (accessed 10-11-15).

Garber, P. 1998. Derivatives in international capital flows. NBER Working Paper 6623, Cambridge, MA: National Bureau of Economic Research, June.

Gereffi, G., J. Humphreys and T. Sturgeon. 2005. The governance of global value chains. *Review of International Political Economy* 12(1): 78–104.

Gereffi, G., and M. Korzeniewicz (eds). 1994. *Commodity Chains and Global Capitalism*. Westport, CT: Praeger.

Germain, R.D. 1997. *The International Organization of Credit: States and Global Finance in the World-Economy*. Cambridge: Cambridge University Press.

Goodhart, Charles. 1981. Problems of monetary management: The UK experience'. In A.S. Courakis (ed.). *Inflation, Depression, and Economic Policy in the West*. Lanham, MD: Rowman & Littlefield, 111–46.

Grinberg, I. 2015. Breaking BEPS: The new international tax diplomacy. Georgetown University Law Centre, 10 September Available at: http://ssrn.com/abstract=2652894.

Hartmann, P. 1998. Currency competition and foreign exchange markets: The dollar, the yen and the euro. Cambridge, UK: Cambridge University Press.

Harvey, D. 1990. *The Condition of Postmodernity: An Enquiry into the Origins of Cultural Change*. Oxford: Blackwell.

Helleiner, E. 1996. *States and the Re-emergence of Global Finance: From Bretton Woods to the 1990s*. Ithaca, NY; London: Cornell University Press.

Helleiner, E., and S. Pagliari. 2011. The end of an era in international financial regulation? A post-crisis research agenda. *International Organization* 65 (1):169–200.

Hu, H.T.C., and B. Black. 2007. Hedge funds, insiders, and the decoupling of economic and voting ownership: Empty voting and hidden (morphable) ownership. *Journal of Corporate Finance* 13:343–67.

IMF. 1993. *Balance of Payments Manual*, 5th edn. Washington, DC: IMF.

IMF. 1997. New international guidelines formulated for recording transactions in financial derivatives. In *Balance of Payments Statistics* (newsletter of the Balance of Payments and Statistics Divisions, IMF) 5(2): December.

IMF. 2005. *Financial Stability Report April 2005*. Available at : https://www. imf.org/External/Pubs/FT/GFSR/2005/01/pdf/chp1.pdf.

Joonkoo, L. 2010. Global commodity chains and global value chains. In *The International Studies Encyclopaedia*, ed.. R.A. Denemark, 2987–3006. Oxford: Wiley Blackwell.

Julius, D. 1990. *Global Companies and Public Policy: The Growing Challenge of Direct Foreign Investment*. London: Pinter.

Kester, A. 1992. *Behind the Numbers: US Trade in the World Economy*, Washington, DC: National Academy Press.

Kester, A. 1995. *Following the Money: US Finance in the World Economy*. Washington, DC: National Academy Press.

Kravis, I., and R. Lipsey. 1992. Sources of competitiveness of the United States and of its multinational firms. *Review of Economics and Statistics*. 74 (2):193–201.

Latour, B. 1987. *Science in Action: How to Follow Scientists and Engineers through Society*. Cambridge, MA: Harvard University Press.

Lewis, M. 2014. *Flash Boys: A Wall Street Revolt*. New York: NY: W.W. Norton.

Leyshon, A. 1996. Dissolving distance?: Money, disembedding and the creation of global financial space. In *P.W. Daniels, and W. Lever (eds). The Global Economy in Transition*. London: Longman.

Leyshon, A., and N. Thrift. 1997. *Money/Space: Geographies of Economic Transformation*. London: Routledge.

MacKenzie, D. 2006. *An Engine, not a Camera: How Financial Models Shape Markets*. Cambridge, MA: MIT Press.

Mackenzie, D. 2011. The credit crisis as a problem in the sociology of knowledge. *American Journal of Sociology* 116 (6):1778–841.

MacKenzie, D., D. Beunza, Y. Millo and J.P. Pardo-Guerra. 2012. Drilling through the Allegheny Mountains: Liquidity, materiality and high-frequency trading. *Journal of Cultural Economy* 5 (3):279–96.

McLeay, M., A. Radia and R. Thomas. (Bank of England Monetary Analysis Directorate) 2014. Money in the modern economy: An introduction. *Bank of England Quarterly Bulletin* Q1 :1–14.

Martin, R. 2002. *Financialization of Daily Life*. Philadelphia: Temple University Press.

Mataloni, R.J. 1995. A guide to BEA statistics on US multinational companies. *Survey of Current Business* March, available at: http://www.bea.gov/scb/account_articles/international/0395iid/maintext.htm (accessed 10-12-15).

Mataloni, R.J. 1998. US multinational companies: operations in 1996. *Survey of Current Business* September, available at: http://www.bea.gov/scb/pdf/INTERNAT/USINVEST/1998/0998mnc.pdf (accessed 10-12-15).

Mehrling, P. 2011. *The New Lombard Street: How the Fed Became the Dealer of Last Resort*. Princeton: Princeton University Press.

Mitchell, T. 2002. *Rule of Experts: Egypt, Techno-politics, Modernity*. Berkeley, CA: University of California Press.

Negri, A. 1988 (1967). Keynes and the capitalist theory of the state post 1929. In *Revolution Retrieved: Selected writings on Marx, Keynes, Capitalist Crisis and New Social Subject 1967–1983*. London: Red Notes, 5–42.

Neilson, J., B. Pritchard and H.W.C. Yeung 2014. Global value chains and global production networks in the changing international political economy: An introduction. *Review of International Political Economy* 21 (1):1–8.

OECD. 1996. *OECD benchmark definition for foreign investmen,*. 3rd edn. Paris: OECD.

OECD. 2013. *Addressing Base Erosion and Profit Shifting.* OECD Publishing. http://dx.doi.org/10.1787/9789264192744-en.

Porter, M. 1990. *The Competitive Advantage of Nations.* London: Macmillan.

Pyrke, M., and J. Allen.2000. Monetized time–space: derivatives – money's 'new imaginary'? *Economy and Society* 29 (2):264–84.

Pryke, M., and R. Lee. 1995. Place your bets: Towards an understanding of globalisation, socio-financial engineering and competition within a financial centre. *Urban Studies* 32 (2):329–44.

Radice, H. 1984. The national economy: A Keynesian myth? *Capital & Class* 8 (1):111–40.

Scholes, M. 1997. Derivatives in a dynamic environment. Nobel Lecture, 9 December. Available at: http://www.nobelprize.org/nobel_prizes/economic-sciences/laureates/1997/scholes-lecture.pdf.

Scott, J. 1976. *The Moral Economy Of The Peasant.* New York, NY: The New Press.

Shamir, R. 2008. The age of responsibilization: on market-embedded morality. *Economy and Society* 37 (1):1–19.

Sharman, Jason. 2010. Offshore and the new international political Economy. .*Review of International Political Economy* 17 (1):1–19.

Sheppard, E. 2002. The spaces and times of globalization: Place, scale, networks, and positionality. *Economic Geography* 78 (3): 307–30.

Taleb, N.N. 2007. *The Black Swan.* New York, NY: Random House.

Teubner, G. 1997. Global Bukowina: legal pluralism in the world society. In Teubner, ed., *Global Law Without a State.* Dartmouth: Ashgate Press.

Thompson, E.P. 1991. *Customs in Common: Studies in Traditional Popular Culture.* New York, NY: The New Press.

Reich, R. 1990. Who is Us? *Harvard Business Review*: January–February.

Reich, R. 1991a. *The Work of Nations: Preparing Ourselves for the 21st Century.* New York, NY: Vintage Books.

Reich, R. 1991b. Rejoinder: Who do we think they are? *The American Prospect*, Winter.

Reich, R. 2005. 1992: As I predicted, only worse. *The American Prospect*: June 2005.

Seabrooke, L., and D. Wigan. 2014. Global wealth chains in the international political economy. *Review of International Political Economy* 21 (1):257–63.

Shiller, R. 2003. *The New Financial Order: Risk in the 21st Century.* Princeton: Princeton University Press.

Tsingou, E. 2015. Club governance and the making of global financial rules. *Review of International Political Economy* 22 (2):225–56.

Tyson, L. 1991. They are not us: Why American ownership still matters. *The American Prospect*: Winter.

Wai, R. 2008. The interlegality of international private law'. *Law and Contemporary Problems* 71: 107–27.

Wyly, E.K., T. Pearce, M. Moos, H. Foxcroft and E. Kabahizi. 2009. Subprime mortgage segmentation in the American urban system. *Tijdschrift voor Economische en Sociale Geografie* 99 (1):3–23.

3

Financial Flows

Spatial Imaginaries of Speculative Circulations

Paul Langley

Introduction: From Chain to Spinning Top

Amidst the recent global financial crisis, 'complexity' was widely posited to be a significant causal factor by practitioners, policymakers and media commentators (Christophers 2009; Datz 2013). By way of simplification and visualization, assertions that global finance was collapsing under the weight of its own complexity were often accompanied by a particular set of maps. Dating from the late 1990s and early 2000s, the maps in question were flow diagrams – typically composed of boxes and arrows, variously coloured and arranged – representing the innovative techniques of asset-backed securitization and their associated array of capital market and derivative instruments. Coming to prominence during the turmoil, these techniques and instruments were supposed to 'complete' and 'perfect' the markets by enabling the widespread distribution of the default risks of subprime mortgage lending, but only succeeded in exacerbating 'systemic risk' (Langley 2008a, 2013). Maps of the so-called 'securitization chain' thus became a ubiquitous feature of the demystifications of the crisis offered by policy and media reports, even featuring in the 2010 Oscar-winning documentary film, *Inside Job*.

As Ismail Erturk and his co-authors suggest, the mapping of this now notorious feature of the global financial market landscape as a chain was significant, not least because the chain was much more than a useful descriptive device for bringing the complexities of global finance down to earth (Erturk et al. 2011). It also 'encouraged the view that securitization

Money and Finance After the Crisis: Critical Thinking for Uncertain Times, First Edition. Edited by Brett Christophers, Andrew Leyshon and Geoff Mann.

was akin to a regular production system that transformed raw inputs – in this case mortgages or other loans – into final products, which were sold on to investors at the end of the production chain' (p. 581). Indeed, the visualization of securitization as a productive and ordered process – 'linear flows with a beginning, middle and end' (p. 582) – had been 'an important influence on light-touch regulatory prescriptions' which assisted financial market innovation prior to the crisis (p. 581).

In place of the flow diagrams of the securitization chain, and taking inspiration from Michel Serres' writings on the appearance of turbulence in the physics of energy and heat flows, Erturk et al. (2011) propose a pair of altogether more illuminating metaphors for envisioning securitization and global financial flows more broadly. For them, the metaphors of 'circuit' and 'spinning top' provide concepts for opening up some of the important dynamics of financial flows. In contrast with a linear chain, thinking in terms of a circuit emphasizes return loops over directional flows; 'everyone gets paid as long as there is throughput in the circuit' (p. 586). And, by adding the metaphor of the child's toy, a spinning top, Erturk et al. underscore the importance of circulatory 'throughput' by reminding us that 'finance requires momentum to remain stable' (p. 586). This is because when flows grind to a halt, as was the case in the crisis, then the edifice of global finance begins to spin off in highly unpredictable and destabilizing directions.

Erturk et al.'s (2011) attention to the role of the securitization chain as an imaginary of global financial flows, and their re-envisioning of securitization techniques as circuit and spinning top, echoes previous interventions by social scientists. Recall, for example, J.K Gibson-Graham's (1996, pp. 135–7) proposal to rethink money and finance as 'the seminal fluid of capitalism', and to thereby stress how the excesses of 'the body of capitalism' periodically become manifest in 'a spasm of uncontrollability and unboundedness'. Gordon Clark (2005), meanwhile, draws our attention to how the metaphorical mapping of global financial markets typically conjures up a relatively smooth space of watery flows. Finance appears as though it is channelled and directed, rather than embedded in uneven geographies which are produced by its inherent tendency to concentrate in key urban centres. For Clark, therefore, it is 'mercury' rather than water that provides an illuminating metaphor for understanding global financial flows, not least because 'mercury pools by its very nature' (p. 104). The ostensible watery qualities of financial flows – creating the impression that markets operate as 'hydraulic systems' that function to irrigate the 'real' economic landscape – also provide the focus for Anne Mayhew's (2011) analysis of the mid-twentieth century development of flow-of-funds accounting by the Federal Reserve. With clear parallels to Erturk et al.'s intervention, Mayhew recovers a debate that turned on whether the movements that practitioner accountants were attempting to capture as a flow-of-funds were more akin to circuits of electricity than to streams and rivers.

Such provocations to interrogate critically the spatial imaginaries of financial flows provide the starting point for this chapter. Here, though, I want to take inquiry in two underexplored directions. First, the chapter asks a different kind of conceptual question from the provocations discussed above. Rather than highlighting that which is elided through an established imaginary (e.g., chain) and proposing a corrective metaphorical concept (e.g., spinning top), I want ask how the generative force of spatial imaginaries might be understood in the context of long-standing conceptual debates that identify speculative circulations as a defining, geographical feature of financial markets. Erturk et al. (2011) allude to such debates when they propose the metaphors of circuit and spinning top, but I bring them into the foreground here. In short, I want to ask how the constitutive contribution of spatial imaginaries to the production and reproduction of speculative circulations can be conceptualized and analysed. Second, I want to draw analytical attention to a trope present within and across prevailing spatial imaginaries that, coming to the surface during the global financial crisis in the USA and UK in particular, envisions financial flows as vital and essential to contemporary life. This trope is politically significant because it helps establish the limits within which debate and disagreement over the future of financial flows takes place. Once speculative circulations appear to be fundamental to securing wealth and wellbeing – despite the insecurity produced by their volatilities and vicissitudes – political debate itself takes on a circular quality in which there appears to be no alternative to what Michel Foucault (2007, p. 49) terms the 'freedom of circulation'.

The chapter advances in three main sections. As the opening section briefly reviews, critical understandings of finance in the contemporary social sciences are marked by an analytical concern with the circulatory and speculative qualities of their problem-object. While they also hold out some promising avenues through which to consider how speculative circulations are actually produced and reproduced, at present they tend not to pay particular analytical attention to the constitutive work which is accomplished by spatial imaginaries. The second section suggests Georg Simmel's century-old contribution to the social theory of money can begin to provide a conceptual basis from which to analyse the constitutive relationships between spatial imaginaries and speculative circulations. The very idea of money and its mobility was crucial to Simmel's (2004) account of the role of money in the making of a particular modern and urban 'style of life' at the turn of the twentieth century. In a similar manner, imaginaries of speculative and uncertain financial circulations can be seen as necessary to the privileged positioning of 'the markets' in the making of a valued form of Anglo-American, neoliberal life during the first decades of the twenty-first century. The third and final section focuses on three particular spatial imaginaries of the speculative circulations of global finance that loomed large in the governance of the recent crisis; namely, the 'liquidity' of money markets, the 'toxicity' of capital markets, and the 'casino' practices of banking.

Speculative Circulations

As post-Keynesian economists Massimo Amato and Luca Fantacci (2012) highlight, the meanings ascribed to 'finance' are often misleading with regard to its temporal–spatial dynamics. The etymology of 'finance' can be traced to the old French terms *fin* (end) and *finer* (to make an end, settle debt), and to the English *fine* (a monetary penalty). Similarly, in contemporary usage the verb 'to finance' usually refers to borrowed monetary resources, especially as they are used to fund purchases and investments. What 'finance' usually invokes, in short, is the two-way flow of a credit–debt relation: credit-money is received, on the one hand, and requires debt (principal and interest) repayment over some period of time, on the other. However, Amato and Fantacci emphasize a notable feature of the temporality of financial relations: when viewed in the aggregate, such relations are never settled, they do not end. Indeed, as the rapid and ongoing expansion of the volume of outstanding household, corporate and sovereign debt in recent decades so vividly attests (Pettifor 2006), the opposite would seem to be case.

Moreover, Amato and Fantacci's (2012) analysis reveals the especially intriguing ways in which the contemporary burgeoning of financial relations without foreseeable end is intimately bound up with the changing spatiality of those relations. Through careful examination of a series of historical developments – most notably, the collapse of the Bretton Woods system of fixed-exchange rates and capital controls – Amato and Fantacci show that the business of wholesale finance has progressively become dominated less by the creation of credit–debt relations in primary markets, and more by the commodification of credit–debt relations and their risk attributes in secondary and derivative markets. In sum, for Amato and Fantacci, contemporary finance without foreseeable end is characterized by the trading of exchangeable and transferable claims upon the future, and by geographies of speculative circulations which are fundamentally different to the two-way flows of a credit–debt relation.

Those writing in the Marxist political economy tradition may well be sceptical of Amato and Fantacci's (2012) claims about the novelty of contemporary developments (e.g., Arrighi 1994), but they would likely agree with their characterization of the temporal–spatial character of finance. Consider, for example, Brett Christophers' (2009) intervention into debates over the complexity of contemporary finance. As he works towards an analysis that takes inspiration from David Harvey's (1982) theoretical development of Marx, Christophers (2009, p. 817) makes the following observation:

> Much of the perceived complexity in modern finance stems ... from the fact that the credit relationship rarely remains, in this day and age, a simple binary one. In other words, it is seldom the case that where money is lent by A to B, *only* A and B remain parties to the money in question: for B will frequently lend that same money on in one way or another, and the loan that A has made to B will *itself* often be passed on in the form of a new credit instrument. Multiple credit instruments and relationships, therefore, but only one underlying pool of money.

In Harvey's (1982) analysis, the credit instruments and relationships that comprise financial flows occupy a position in the circuits of capitalism which is, at once, essential and destabilizing. He posits, for instance, that 'the circulation of interest-bearing capital performs certain vital functions and the accumulation of capital therefore *requires* that money capitalists achieve and actively assert themselves as a power external to and independent of actual production processes' (p. 261, original emphasis). There are, for Harvey, six such 'functions' (pp. 262–72). Banking and finance: (1) mobilize money that would otherwise lay idle; (2) provide credit that accelerates trade and the circulation of commodities; (3) increase the rate of commodity production through fixed investment; (4) facilitate the formation of 'fictitious capital' (wherein credit is extended indefinitely as resaleable ownership claims that are always waiting to be realized, i.e., stocks and shares); (5) encourage the equalization of profit rates across the economy because of 'flows in response to profit rate differentials' (p. 270), and; (6) embolden the dynamism of capitalism by assisting the centralization and restructuring of capital. In return for the performance of these vital functions, banking and finance extract rent in the form of the appropriation of interest. In the first instance and in aggregate, then, the movements and machinations of finance capital under capitalism are understood primarily in terms of the two-way flows of the power relation of credit–debt.

At the same time, a feature of the theoretical development of Marx offered by Harvey (1982) is that pre-capitalist tendencies for money lending to be largely usurious and parasitical – tendencies 'which Marx was aware of but brushes aside' (p. 257) – continue to be present in his account of the place of finance capital within capitalism. When understanding finance capital as 'a process, as opposed to a particular set of institutional arrangements or a catalogue of who is dominating whom within the bourgeoisie', Harvey highlights a number of 'circumstances' and 'barriers' that prevent it from 'functioning as a fine-tuner of accumulation' (p. 287). These include the persistent tendency for money capital to be 'indiscriminate as to its uses since it typically flows to appropriate revenues of no matter what sort' (p. 287). Although the two-way circulation of interest-bearing capital is thus crucial to the disciplinary power and dynamism of the credit–debt relation in capitalism (see also Lazzarato 2015), for Harvey it remains the case that 'there is nothing to prevent speculative investment in the appropriation of revenues from getting entirely out of hand' (p. 287). The result is that capitalism is characterized not only by a 'surface froth of perpetual speculation' (p. 325), but also by tendencies to 'speculative fever' which periodically come to dominate as a discernible phase in the ups-and-downs of the 'accumulation cycle' (p. 303). The speculative passions and flows of finance capital that Harvey identifies are not, however, a simple destabilizing force within capitalism. As refined in his later work, speculative circuits are understood by Harvey (2001) to be capable of providing a 'spatio-temporal fix' for the contradictions of capitalism. By switching excess capital into secondary circuits and establishing claims upon specific urban property markets during a particular conjuncture, for example, finance capital is able to paper over the cracks of crisis-prone capitalism. Such fixes remain necessarily temporary, however, as they also serve to widen the reach and deepen the intensity of capitalism's destabilizing contradictions.

The post-Keynesian and Marxist political economy traditions can thus be seen to leave us in little doubt: aggregate financial flows are not merely the cumulative result of two-party credit–debt relations, but necessarily also include the speculative circulations of transferable claims on the future. Put baldly, much of the flow of finance which passes as necessary for the operation of the 'real economy' of production and consumption – and, increasingly, for the provision of the 'public goods' of urban infrastructure (e.g., transportation, energy, waste, communication) (Allen and Pryke 2013; O'Neill 2010, this volume), and even the preservation of liveable environments for future generations (Sullivan 2012) – is actually speculative circulations.

Furthermore, as it pinpoints the speculative quality of financial flows with considerable clarity, Marxist political economy also holds that the conditions of possibility for speculative circulations are systemic to capitalism. To paraphrase Harvey (1982, p. 261), speculative circulations are an inevitable consequence of the 'power external to and independent of actual production processes' that 'money capitalists' are required to 'achieve and actively assert'. It follows that the contemporary pervasiveness of the speculative circulations which encircle the globe is largely a consequence of the particular achievements and assertions of power by finance capital, in the USA in particular (Gowan 1999; Panitch and Gindin 2013). However, if we look beyond Marxist political economy to the plural, critical understandings of finance which are also present in the social sciences, we find that analytical emphasis is typically placed upon an array of contingent conditions that might also be said to make speculative circulations possible. A number of conceptual routes for analysing the processes and practices that produce speculative circulations are opened up, for example, by the recent calls to draw together cultural economy concepts with over two decades worth of research into the geographies of money and finance (Hall 2011; Langley 2016).

Reflecting a core concern with how relatively discrete and variegated financial markets are assembled through socio-technical and material processes (MacKenzie 2009), what cultural economy suggests is that economic formulas, models and other calculative devices play a crucial role in making the more-or-less discrete appropriations of speculative circulation possible. This may help put some flesh on the bones of Maurizio Lazzarato's (2015, p. 43) recent claim that rent – as an apparatus that appropriates and captures value in ways distinct from the apparatuses of profit and taxation – is an 'accounting machine' that includes its own 'criteria for evaluation and comparison, its own measurements, and its own property regime'. And, if the dynamic category of 'value' is held to be the foundational and 'organising principle of capital' (Mann 2010a, p. 175), then it follows that recognizing the contingent constitutive force of economic formulas, models and devices is critical to explaining how the rent-like appropriations of speculative circulations come to figure so positively in the prevailing value regime (Cooper and Konings 2015). At a minimum, while post-Keynesian and Marxist political economy foreground the speculative qualities of financial flows, cultural economy accounts of financial markets certainly help to deepen understanding of the contingent socio-technical and material conditions which make speculative circulations possible.

If we take analytical insights from geographic research into the significance of financial centres, meanwhile, there is clearly also a sense in which the conditions and processes that sustain speculative circulations are place-based. Once financial centres are understood as crucibles of financial information, trust and knowledge/power (Clark 2005; Lee 2011; Thrift 1994), then the speculative circulations of finance can be understood to be manufactured, directed and made meaningful through calculative and interpretative practices in such centres. And, as ethnographic research into the workplaces of the City of London and Wall

Street highlights (Ho 2009; McDowell 1997), the production of speculative circulations can also be understood as an embodied process, requiring the calling-forth and performance of particular subjectivities. In the contemporary City, for example, it is clear that much has changed from the era of 'gentlemanly capitalism' (Augar 2000), but it remains unclear at present what the recent crisis means for the exuberant and aggressive masculine subjectivity of the market trader which was highly valorized across recent decades (Hall and Appleyard 2012). This is especially the case given the rise of high-frequency trading and the role of algorithms in trading strategies (MacKenzie et al. 2012), and the associated valorization of the embodied mathematical knowledge of so-called 'quants' (Patterson 2011). The present moment would seem to require a more thorough questioning of the embodied processes through which speculative circulations become possible (Borch, Hansen and Lange 2015), particularly as they unfold in financial centres and workplaces.

What this brief review of critical social scientific research makes plain, in sum, is that speculative circulations are an inherent feature of capitalist finance, and that the analysis of how these circulations are made possible in different contexts should consider configurations of class and sovereign power and the complex concentrations of economic knowledge and gendered embodied work which take place in global financial centres. It remains the case, however, that the knowledge and practices of finance which tend to be addressed by existing research are largely confined to expert knowledge and professional practices. There is limited attention to the highly-mediated perceptions and dispositions concerning the uses and circulations of finance that travel and spread throughout contemporary societies (de Goede 2012). As Urs Stäheli (2013) argues, with reference to the USA between the 1870s and 1930s – a context in which financial markets first began to take on enormous importance in mainstream culture – the spectacle of speculation is translated into popular knowledges and discourses in ways that are crucial to sustaining and legitimating its circulations. This is arguably intensified in the contemporary period, when the fluctuations and undulations of 'the markets' have become the basic feedstock of a 'finance–entertainment complex' operating in real-time (Taylor 2004, p. 191; Clark, Thrift and Tickell 2004). Although it places the speculative qualities of financial flows to the front-and-centre of analysis and provides significant insights for exploring their constitution, then, critical social science presently gives little analytical weight to the force and play of spatial imaginaries when accounting for the production and reproduction of speculative circulations. In the following section, I turn to the social theory of money to begin to provide a conceptual basis from which to analyse the constitutive relationships between spatial imaginaries and speculative circulations.

Spatial Imaginaries

Originally published in 1907, Georg Simmel's (2004) *The Philosophy of Money* is widely regarded as a principal contribution to the classical social theory of money. The prevailing contemporary reading of this text follows Zelizer's (1994) influential critique and positions Simmel – alongside Marx, Weber and others – as a theorist who emphasizes money's calculative quality as a means of accounting that renders things commensurate (e.g., Gilbert 2005). For Simmel, it is the circulation of the quantifications of money which are crucial to understanding how 'a coherent world of value' – where value appears as an objective property of objects themselves – can emerge through the interplay of 'subjective desire' and 'intersubjective processes of exchange' (Dodd 2014, pp. 27–9). Yet, the operation of money is, for Simmel, not solely understood as the circulation of a pure instrument of quantitative valuation (Dodd 1994, 2014). Rather, the circulation of money also carries with it the very idea of money; that is, the symbolic meanings and associations necessarily attached to money. It is in these terms that Simmel probes the relations between money in circulation and the 'individual freedoms' of a modern, metropolitan 'style of life' that was emerging at the turn of the twentieth century. By 'uniting people while excluding everything personal and specific' (Simmel 2004, p. 347), both monetary calculation and the idea of money are crucial to the modern loosening of social bonds and flattening of social hierarchies, as all things and desires come to appear to be available at a price.

For John Allen and Michael Pryke (1999), the emphasis Simmel places on the idea of money and its contribution to modern social life provides a critical vantage point from which to understand how the time–spaces of contemporary global financial flows are produced and reproduced. To extend Simmel's insights, Allen and Pryke make two principal conceptual moves in the context of classical and contemporary writings on the social theory of money. First, while Simmel devoted his analytical attention somewhat narrowly to the circulation of money as currency in a world of exchange, Allen and Pryke conceive money and finance more broadly. This follows from a number of different perspectives found in the contemporary social theory of money. Here 'money', in the first instance, is a means of accounting and sustaining the uncertain promises to pay of credit–debt relations rather than an enabler of exchange (Ingham 2004; Lazzarato 2012). The claims and obligations of credit–debt themselves become transferable, and can be represented as financial instruments to be speculated upon in the markets, because they are denominated in a shared money of account.

Second, Allen and Pryke (1999) elaborate upon Simmel's concerns with the idea of money through the categories of 'money imaginary' and 'cultures of money' (see Dodd 1994). While Simmel sought to understand how the idea of money produced a particular 'style of life' somewhat specific to metropolitan Berlin, for Allen and Pryke there are multiple imaginaries that make the contemporary movements of money and finance meaningful and which contribute to multiple money cultures. As they have it, with derivatives in particular, 'it is not the instruments so much that are of concern', but 'the ideas about what they may facilitate and what different groups of people in locations distant from one another imagine themselves to be involved in' (Allen and Pryke 1999, p. 52). It follows that the shared meanings of 'money cultures' are 'lived, experienced, and interpreted by particular groups in particular places at particular times', and are 'made up of people who position themselves in relation to the circulation of money and are also positioned by it' (p. 65). What specific monetary and financial circulations are taken to mean thus influences not only actions and responses, but the formation of subjectivities (see Langley and Leyshon 2012).

Allen and Pryke's (1999) second conceptual move runs somewhat counter to that which is typically made in the contemporary social theory of money. Following Zelizer's (1994) intervention (see Dodd 2014, pp. 286–94), monies in use and circulating alongside of each other are typically held to be acted on and 'ear-marked'. In short, according to the contemporary social theory of money, the circulation of money is not simply a corrosive force that wipes away social bonds. Rather, it is said to be subject to cultural pressures and processes which can ensure that, when it is in use, money is employed in ways that are largely consistent with already existing identities and interests. For Allen and Pryke (1999, p. 65), in contrast, there can be 'no endless play of money cultures'. This is because, grounded in Simmel's assertion that the circulation of money also carries with it the very idea of money, money is always already a spatial imaginary and a culture. It is not a thing that, empty of meaning at the outset, can be culturally or politically ear-marked when put to use. There may well be a 'diversity of cultural interpretations related to movement and mobility' which are of constitutive significance to the circulations of monies (p. 65). But, amidst this diversity are prevailing imaginaries that, stressing the crucial role of money and finance in modern social relationships, hold a particular and pervasive generative force that contributes to sustaining monetary and financial circulations and summoning up monetary and financial subjectivities.

Rooted in Simmel's concern with the symbolic meanings and associations always already attached to modern money, and extended through Allen and Pryke's (1999) prescient contribution, what can be said of how spatial imaginaries provide conditions of possibility for speculative circulations? The primary focus for Allen and Pryke's (1999) own analysis is derivative markets, and thus upon how imaginaries of the mobility of money in those markets give meaning to elite, professional practices in the financial centres where these markets are located. As they have it, the symbolic marking of derivative circulations by their pace and simultaneity 'positions many of those working in finance in a culture of fast risk and effortless gain' (p. 62). Allen and Pryke (1999) are also careful to differentiate the particular culture of derivatives markets from the popular culture of global finance that, they suggest, has taken hold in recent decades and provides 'a wider circuit of meaning' for 'personal finance' in the USA and UK in particular (p. 64). As they have it, quoting terms from Simmel, today's popular culture of finance features a spatial imaginary that is characterized by 'restless flow' and 'effortless multiplication', such that 'a world of derivatives and speculative gain' appears to be 'not restricted to a handful of dealers in the world's financial centres' (p. 64).

The presence and play of spatial imaginaries of speculative circulations in everyday life which Allen and Pryke (1999) highlight can be understood, moreover, as a manifestation of the biopolitical mode of power and security that Foucault (2007, 2008) identified in his later work (Langley 2015). For Foucault (2007), one of the features that distinguishes biopolitical power from other modalities (i.e., sovereignty and discipline) is the way in which it configures and seeks to act upon circulations. Circulations of 'people, merchandise, and air etcetera' may well be uncertain and pose dangers (p. 29), but it is apparently upon these very circulations and their contingencies that the dynamic and entrepreneurial production of wealth, health and happiness rests (Foucault 2008). With a clear parallel to Simmel's account of the role of the idea of money in the making of a modern and urban 'style of life' at the turn of the twentieth century, on these terms the last three to four decades have seen uncertain financial market circulations come to be understood as vital to securing a valued form of neoliberal life.

As constituted through the prevailing spatial imaginary, the ostensible significance of speculative circulations to contemporary socio-economic life does not simply turn on their growing contribution to economic growth (Christophers 2013), or on finance's continuously asserted role of providing investment that is said to nourish the 'real economy' of production and employment. Rather, what is especially notable is how this spatial imaginary figures the close interrelations that have developed between the saving and borrowing of everyday life in Anglo-America, on the one hand, and the speculative circulations of global financial markets, on the other (Langley 2008b). The rise of mutual fund stock market investment has increasingly displaced retail deposit saving and collective retirement insurance. In consequence, security seems to turn less than previously upon thrifty provision for the future and the calculation, pooling and spreading of uncertainties as 'risks'. Instead, it now appears to turn on the entrepreneurial embrace of the risks and uncertainties of financial circulations, accompanied by the depoliticization of the unequal distribution of rewards. At the same time, relatively unencumbered access to mortgage and consumer credit would now seem to play a positive role in facilitating the prosperity of all. The result, as Lazzarato (2012, p. 94) argues, is that the management of debt – alongside the management of precarious employability, falls in real wages, and the shrinking availability of public services – characterizes the lived experience of (in)security for the majority. But, imagined as vital and essential to contemporary life, finance and its speculations remain free-flowing and largely unencumbered by the kinds of restrictions and regulations that were in place four decades ago.

The Governance of the Global Financial Crisis

One method for analysing how highly-mediated spatial imaginaries envision speculative circulations as essential to contemporary life is to interrogate how these imaginaries come to the surface during moments of crisis, featuring in particular in the interventions that seek to render and govern crises. This is clearly a somewhat different form of inquiry to that which, inspired by Marxist political economy, posits the 'vital functions' (Harvey 1982) of financial flows within capitalism and seeks to explore various attempts by central banks, treasuries and regulators aimed at restoring those circulations (e.g., Epstein and Wolfson 2013). For Marxist political economy, the spatial imaginaries of financial flows which are present in crisis management would not seem likely to be accorded analytical weight as generative forces in their own right. Yet, one of the notable features of crises is that they are not governed as crises of finance and capitalism, *tout court*, but through relatively discrete technical problematizations in which the rendering of issues of circulation looms large (Langley 2015). At a general level, for example, crises of banking and finance are typically defined in economic theory and practice as 'liquidity crises' (Allen et al. 2011). Whether finding form as a 'run' by depositors on a fractional reserve bank, for instance, or as a rupture in money and capital market flows, it is a 'liquidity squeeze' that is typically held to be the essential and defining feature of a financial crisis.

To take a more specific example, consider the spatial imaginary that prevailed in the governance of the Asian financial crisis of 1997–98. Here the technical problem that was fabricated and managed was 'international capital mobility'. This spatial imaginary was crucial to the governance of the crisis by the International Monetary Fund (IMF) and other international financial institutions (IFIs). State-imposed capital controls to restrict disruptive cross-border flows, or transaction taxes to slow capital flows within and between markets, were both eschewed by the IFIs on the grounds that such measures interrupted the capacity of global finance to provide vital flows for 'real' economic investment (Watson 2007; Gallagher 2015). What emerged from the crisis for the IFIs was a developmental agenda that called for the more careful 'sequencing' of financial reform, wherein the deregulation of domestic financial systems preceded the removal of capital controls which remained necessary to enable the benefits of free-flowing finance (Langley 2004; see Lai and Daniels, this volume).

In the governance of the global financial crisis that began in the autumn of 2007, meanwhile, the figuring of the technical problems to be addressed featured an array of spatial imaginaries of speculative circulations. In the Anglo-American heartland of the crisis in particular, these imaginaries were notable for the ways in which they construed anything less than the full restoration of circulations as a fundamental threat to the security of the population. What was to be secured through crisis governance was not merely the markets, the banks, and the financing of the 'real economy' they are said to provide for, but popular stock market investment and privatized pensions, on the one hand, and expanded and widespread availability of mortgage loans and consumer credit, on the other. Restoring the uncertain circulations of finance and banking was figured, in short, as a matter of security, an urgent need to refurbish a particular and valued form of Anglo-American, neoliberal life. Contrary to Harvey (2011, p. 7), it was not that 'everyone in power recognized' the crisis as 'a matter of life and death for capital', but that there was a tendency for the crisis to be rationalized and strategized as a threat to the security of life itself.

Take, for example, how during the first twelve months or so the crisis was made legible and governed as a problem of 'liquidity'. As noted above, crises of banking and finance are typically defined in economic theory and practice as 'liquidity crises'. As such, it was no surprise when the media consistently iterated the global financial crisis as a situation in which liquidity had 'evaporated' from markets which were also said to have 'dried up', 'seized up' or 'frozen over'; or that the Federal Reserve and Bank of England enacted complex and unprecedented last resort lending interventions which were cast as 'pumping' or 'injecting liquidity'. The markets in question were the money and capital markets: in money markets, banks and other institutions were unable to access short-term debt or renew ('roll-over') outstanding obligations; in capital markets, there were few buyers, as institutions sought to offload vast portfolios of assets related to and derived from subprime mortgages. That the purchase of capital market assets had tended to be funded through the money markets only compounded the issue.

Yet, governing the crisis as a technical problem of liquidity also had a number of wider symbolic and cultural resonances. To recall the provocations of Clark (2005) and Mayhew (2011) noted at the outset, the watery spatial imaginary of liquidity conjures the circulations of finance in a particular way: markets appear as functioning to provide the flows essential to the irrigation of the 'real' economic landscape. Prior to the crisis, the spatial imaginary of liquidity had become something of a totem for the innovations that had supposedly 'completed' and 'perfected' the markets (Cooper 2014). As Pasanek and Polillo (2011, pp. 232–3) note, 'liquidity' borrows from broader representations of markets which gives 'flows of currency and credit the appearance of a natural and automatic process', and thereby 'reinforces laissez-faire models of equilibrium and circulation'. As such, when the crisis hit, the liquidity of markets appeared as 'the prerequisite for the very possibility of credit' (Amato and Fantacci 2012, p. 24). It followed from this spatial imaginary of financial circulation that the so-called 'credit crunch' acutely experienced by firms and domestic borrowers in retail markets was indeed a direct consequence of the illiquidity prevailing in money and capital markets. Put another way, as Bryan and Rafferty (2013) have it, if the markets could not provide the liquidity to keep the circulations of finance in motion, then it appeared that central banks had little choice but to rip up their rulebooks to provide liquidity as a 'public good'.

The first year or so of the crisis also saw the enactment of a number of governance initiatives, in the USA in particular, that turned on a spatial imaginary that envisioned the speculative circulations of the capital markets in a related but somewhat different manner. When the Federal Reserve acted to save Bear Stearns in March 2008 and American International Group (AIG) in November 2008, and when the US Treasury unveiled the Troubled Assets Relief Programme (TARP) at the height of the crisis in September 2008, the technical problem to be governed was said to be one of 'toxicity'. Subprime assets, now illiquid and thus impossible to value, were commonly referred to as 'toxic assets', 'toxic refuse', and 'toxic waste'. This marked something of a departure from previously prevailing descriptions of failing assets as 'junk', and as that which could be merely discarded without consequence. Assets that were 'toxic', in contrast, were poisonous to those that held them. The supposed solution was to create 'bad banks'. This was a form of crisis management that drew on two previous crisis governance experiments: the 1989 Resolution Trust Corporation, created in response to the US Saving and

Loans debacle; and the aptly named Retriva and Securum bad banks that were established in 1992 in response to the Swedish banking collapse. The aim of the bad banks created in the global financial crisis was to take toxic assets temporarily out of circulation, thereby making it possible for markets to return to the norm of price discovery necessary for the circulation of transferable claims.

Talk of toxicity replaced the hydraulic and watery qualities of financial flows with an altogether more embodied spatial imaginary of circulation. The description of assets as 'toxic' called-up the long-standing representation of money and finance as the blood of the economy-body (Johnson 1966; Mann 2010b; Simmel 2004, p. 474). It thus appeared that the cardiovascular financial system of the economy-body had been momentarily poisoned, a vision drawing sustenance from the medical etymology of the term 'crisis' and the related tendency to represent the financial crisis itself through an array of analogies with contagious diseases (Peckham 2013). It followed that the necessary solution was a set of 'bad bank' interventions akin to a purification or dialysis of the financial bloodstream. What was also notable, moreover, were the various ways in which siphoning-off toxic assets was explicitly represented as a move which would prevent the infection from spreading to the rest of the economy, endangering life itself. For example, when explaining Bear Stearns' bad bank to a Senate committee in April 2008, then-president of the Federal Reserve Bank of New York, Timothy Geithner (2008), stated that 'if this dynamic continues unabated' without 'a forceful policy response', then 'the consequences' would include 'higher borrowing costs for housing, education, and the expenses of everyday life' and 'lower value of retirement savings'. Similarly, in the televised address to the nation he delivered when the TARP proposals were before Congress, President George W. Bush (2008) stressed that 'This rescue effort is not aimed at preserving any individual company or industry. It is aimed at preserving America's overall economy.' For Bush, it was clear that 'our entire economy is in danger', and that the costs of inaction would be high:

> without immediate action by Congress, America could slip into a financial panic and a distressing scenario would unfold. More banks could fail ... The stock market would drop even more, which would reduce the value of your retirement account. The value of your home could plummet ... Even if you have good credit history, it would be more difficult for you to get the loans you need to buy a car or send your children to college.

Bush's televised address left viewers in little doubt that toxic assets were a problem for the American way of life, and that the TARP provided the solution to that problem. As he later put it (in Politi 2008), the TARP was required 'for the financial security of every American'.

As the crisis deepened in the autumn of 2008, the technical and circulatory problem to be addressed by crisis management was rendered differently yet again. Now it seemed that the critical issue was the solvency of banking which, in popular parlance, was 'underwater' and in need of 'bail-out'. The action apparently required to restore the solvency of banking thus appeared to be 'recapitalization', and not the 'nationalization' that these actions were also recogniszd as carrying forward. Enacted through various sovereign treasury programmes on either side of the Atlantic during 2008 and 2009, the bail-outs were figured through a spatial imaginary that envisioned the credit flows of commercial banking as particularly crucial to contemporary, Anglo-American life (Langley 2015, pp. 97–9). Of particular interest to us here, however, is the content of the ensuing interventions which centred on the regulation of banking. The problem in this respect was widely held to be that so-called 'permissive', 'lax' and 'light-touch' regulation prior to the crisis had enabled commercial banks to become 'too big to fail' through excessive risk-taking in wholesale markets. In the terms that travelled through political and media circles in the UK and in the wake of the bail-out of Royal Bank of Scotland in particular, banks had not been concentrating on their 'core business' of lending to businesses and families, and instead had become embroiled in 'the casino' of speculative circulations (e.g., Turner 2009, p. 94; *Guardian* 2010).

A spatial imaginary that differentiates between types of financial flows – contrasting the apparently steady saving and lending of commercial banking with the reckless and rash gambling of the markets – was also a feature of US regulatory responses to the Wall Street Crash of 1929, most notably the Banking Act (known as the Glass–Steagall Act) of 1933. Given impetus by the Pecora Hearings, the Banking Act of 1933 was carried through Congress by former treasury secretary, Carter Glass, and then-chairman of the House Banking and Currency Committee, Henry B. Steagall. While it also created the Federal Deposit Insurance Corporation, the legislation was especially significant in the way that it condemned universal banking and firmly divided Main Street's deposit-taking commercial banks from Wall Street's investment banks. Glass in particular had become the spokesman for a period in popular culture marked by revulsion at the so-called 'madhouse' of Wall Street during the roaring 'twenties (Fraser 2005, p. 367). As President Franklin D. Roosevelt famously put it during his inauguration speech in March 1933, 'Practices of the unscrupulous money changers stand indicted in the court of public opinion, rejected by the hearts and minds of men' (in Chancellor 2000, p. 220). The spatial imaginary that differentiated between financial flows in the US in the wake of the Wall Street Crash was thus one in which the speculative circulations of stock markets posed a security threat to valued life, and accordingly the excesses of the markets were to be curtailed and prohibited.

The headline regulatory interventions in the USA and UK in the recent global financial crisis – namely, the 'Volcker rule' included in the Dodd–Frank Wall Street Reform and Consumer Protection Act of late 2010, and the 'Vickers' ring-fence' which, emerging from the Independent Commission on Banking report in 2011, featured in the Banking Reform Act of 2013 – were certainly similar to the Glass–Steagall Act. Indeed, Paul Volcker, the former head of the Board of Governors of the Federal Reserve who was the driving force behind the regulation that bears his name, certainly took inspiration from Carter Glass, and the Volcker rule was often referred to in media and policy circles as 'Glass–Steagall lite'. Yet, as they sought to specify the permissible practices of banks across the domains of financial flows that they differentiated, the Volcker rule and the Vicker's ring-fence did not condemn and curtail speculative circulations. Rather, certain wholesale market circulations deemed high-risk and casino-like were positioned at one remove from deposit-taking banks. While the flow of credit from commercial banking thus appeared to provide the financial essentials of life, speculative circulations nonetheless continued to be imagined as crucial to securing wealth and well-being.

Conclusions

Money moves, finance flows. And, as stressed by classical and contemporary social scientific research conducted from a wide range of perspectives, circulation is crucial to the very nature and existence of money and finance. As Georg Simmel asserts (2004, p. 571), for example, 'when money stands still, it is no longer money according to its specific value and significance'. It is clearly no surprise, then, that money and finance are replete with popular terminology, figures of speech, metaphors and maps that furnish these circulatory geographies with spatial imaginaries. To date, however, research which draws attention to the significance of representations of the movements of money and finance concentrates on how they obscure certain underlying or arguably more telling dynamics. With reference to financial flows specifically, the task for critical analysis would seem to be one of eschewing prevailing imaginaries – especially as they suggest productive inflows into the 'real economy' – and of proposing an alternative envisioning capable of bringing the uneven, volatile and/or speculative qualities of circulations to the fore.

In contrast, and in the context of previous social science that underlines that the geographies of financial markets are marked by speculative circulations, this chapter's first motivation is to ask how the generative force of spatial imaginaries might be understood in conceptual terms. While extant research opens up a number of conceptual routes for considering the production and reproduction of speculative circulations – from class and state power to the complex concentrations of embodied work, knowledge and calculation in global financial centres, for example – it largely neglects the role of spatial imaginaries. The chapter takes inspiration from Georg Simmel's (2004) attention to the symbolic meanings and associations necessarily attached to money in circulation, and to how the very idea of money is crucial to the making of a particular modern and urban 'style of life'. From Allen and Pryke's (1999) extension of Simmel's concepts in their analysis of derivatives markets, and through recourse to Foucault's (2007, 2008) account of the biopolitical mode of power and security, the chapter offers an approach for understanding how prevailing spatial imaginaries contribute to the constitution of contemporary speculative circulations.

The second motivation for the chapter is analytical and political in character. The omission of the generative force of spatial imaginaries from critical understandings of speculative circulations certainly presents a conceptual problem. But this omission is perhaps only of wider significance because of the highly-mediated representations of 'the markets' and their mobilities that, especially in present-day popular culture in the USA and UK, envision financial flows as vital and essential to the contemporary, neoliberal style of life. The analysis offered here illuminates and explores the presence of spatial imaginaries of speculative circulations in the rendering and governing of the recent global financial crisis. This includes how the crisis was made-up and managed as the technical problems of 'liquidity' in money markets, 'toxicity' in capital markets, and the regulation of 'casino' practices in banking. What this shows, moreover, is that the spatial imaginaries of financial flows at work in the governance of the global financial crisis did not necessarily deny the speculative character of circulation. The markets and banks that were being saved from themselves through unprecedented liquidity injections, bad banks, bail-outs and the like were not solely envisioned as the source of flows necessary for the operation of the 'real economy'. Speculative circulations were also deeply implicated in securing the wealth and well-being of the population because of the entrepreneurial opportunities they apparently afford.

Giving conceptual and analytical consideration to the constitution of speculative circulations through spatial imaginaries thus sheds some fresh light on what is one of the principal and pressing political questions provoked by the recent global financial crisis: why, despite the magnitude of the ruptures experienced during the crisis, has global finance emerged largely unscathed, with its speculative circulatory excesses intact? Or, as French and Leyshon (2010) put it, why has the crisis proved to be such a terrible waste for progressive politics? This is not to claim, of course, that existing critical social scientific research into money and finance does not also provide useful vantage points from which to consider this question. It would indeed seem fair to conclude that class and state power have been instrumental in ensuring that global finance remains very much open for business, and that the expert knowledge networks centred on Wall Street and the City of London have largely continued to define the 'anti-political' (Barry 2002) calculative market terrain upon which debate takes place. What recognizing the generative force of spatial imaginaries suggests, however, is a somewhat different conclusion. It is also a conclusion that is, arguably, a more disturbing one. Prevailing and highly-mediated spatial imaginaries stymie popular debate and disagreement over the freedoms of speculative circulations in what are typically reified as 'the markets' (see also Christophers 2015). In doing so, they also forestall the emergence of what William Connolly (2002) calls a genuinely pluralist and ethical politics, and a society in which asset holders and prime borrowers would be willing to call into question their own identities and the personal security that speculative circulations supposedly produce.

References

Allen, F., E. Carletti, J.P. Krahnen and M. Tyrell (eds). 2011. *Liquidity and Crises*. New York: Oxford University Press.

Allen, J., and M. Pryke. 1999. Money cultures after Georg Simmel: Mobility, movement, and identity'. *Environment and Planning D: Society and Space*. 17 (1):51–68.

Allen, J., and M. Pryke. 2013. Financializing household water: Thames water, MEIF, and 'ring-fenced' politics. *Cambridge Journal of Regions, Economy and Society* 6 (3): 19–439.

Amato, M., and L. Fantacci. 2012. *The End of Finance*. Cambridge: Polity Press.

Arrighi, G. 1994. *The Long Twentieth Century: Money, Power and the Origins of Our Times*. London: Verso.

Augar, P. 2000. *The Death of Gentlemanly Capitalism: The Rise and Fall of London's Investment Banks*. London: Penguin.

Barry, A. 2002. The anti-political economy. *Economy and Society* 31 (2):268–84.

Borch, C., K.B. Hansen and A-C. Lange. 2015. Markets, bodies and rhythms: A rhythmanalysis of financial markets from open-outcry trading to high-frequency trading. *Environment and Planning D: Society and Space* 33 (6):1080–97.

Bryan, D., and M. Rafferty. 2013. Fundamental value: A category in transformation. *Economy and Society* 42 (1):130–53.

Bush, G.W. 2008. Transcript: President Bush's speech to the nation on the economic crisis. *New York Times*, 24 September, viewed 19 May 2011, http://www.nytimes.com.

Chancellor, E. 2000. *Devil Take the Hindmost: A History of Financial Speculation*. Basingstoke: Macmillan.

Christophers, B. 2009. Complexity, finance, and progress in human geography. *Progress in Human Geography* 33 (6):807–24.

Christophers, B. 2013. *Banking across Boundaries: Placing Finance in Capitalism*. Chichester: John Wiley & Sons Ltd.

Christophers, B. 2015. Against (the idea of) financial markets. *Geoforum* 66 (1):83–93.

Clark, G.L. 2005. Money flows like mercury: The geography of global finance. *Geografiska Annaler* 87B (2):99–112.

Clark, G.L., N. Thrift and A. Tickell. 2004. Performing finance: The industry, the media and its image. *Review of International Political Economy* 11 (2):289–310.

Connolly, W.E. 2002. *Identity/Difference: Democratic Negotiations of Political Paradox*. Minneapolis, MN: University of Minnesota Press.

Cooper, M. 2014. The strategy of default: Liquid foundations in the house of finance. *Polygraph: An International Journal of Culture and Politics* 33 (24):79–96.

Cooper, M., and M. Konings. 2015. Contingency and foundation: Rethinking money, debt and finance after the crisis. *The South Atlantic Quarterly* 114 (2):239–50.

Datz, C. 2013. The narrative of complexity in the crisis of finance: Epistemological challenge and macroprudential policy response. *New Political Economy* 18 (4):459–79.

de Goede, M. 2012. *Speculative Security: The Politics of Pursuing Terrorist Monies*. Minneapolis, MN: University of Minnesota Press.

Dodd, N. 1994. *The Sociology of Money: Economics, Reason and Contemporary Society*. Cambridge, UK: Polity Press.

Dodd, N. 2014. *The Social Life of Money*. Princeton, NJ: Princeton University Press.

Epstein, G.A. and M.H. Wolfson. 2013. Introduction: The political economy of financial crises. In *The Handbook of the Political Economy of Financial Crises*, eds M.H. Wolfson and G.A. Epstein, 1–20. Oxford: Oxford University Press.

Erturk, I., J. Froud, A. Lever and K. Williams. 2011. Changing the metaphor: finance as circuit. *Socio-Economic Review* 9 (3):580–8.

Foucault, M. 2007. *Security, Territory, Population: Lectures at the Collège de France, 1977–1978*, trans. G. Burchell. Basingstoke: Palgrave Macmillan.

Foucault, M. 2008. *The Birth of Biopolitics: Lectures at the Collège de France, 1978–1979*, ed. M. Senellart, trans. G. Burchell. Basingstoke: Palgrave Macmillan.

Fraser, S. 2005. *Wall Street: A Cultural History*. London: Faber & Faber.

French, S., and A. Leyshon. 2010. 'These f@#king guys': The terrible waste of a good crisis. *Environment and Planning A* 42 (11):2549–59.

Gallagher, K.P. 2015. Contesting the governance of capital flows at the IMF. *Governance: An International Journal of Policy, Administration, and Institutions* 28 (2):185–8.

Geithner, T. 2008. Actions by the New York Fed in response to liquidity pressures in financial markets. Committee on Banking, Housing and Urban Affairs, US Senate, Washington, DC, 3 April, viewed 13 November 2011, http://www.newyorkfed.org/newsevents/speeches/2008/gei080403.html.

Gibson-Graham, J.K. 1996. *The End of Capitalism (as we knew it): A Feminist Critique of Political Economy*. Oxford: Blackwell.

Gilbert, E. 2005. Common cents: Situating money in time and place. *Economy and Society* 34 (3):357–88.

Gowan, P. 1999. *The Global Gamble: Washington's Faustian Bid for World Dominance*. London: Verso.

Guardian. 2010. Barclays' Bob Diamond hits out at criticism of 'casino banks'. *Guardian*, 12 September, viewed 10 December 2012, http://www.theguardian.com.

Hall, S. 2011. Geographies of money and finance I: Cultural economy, politics and place. *Progress in Human Geography* 35 (2):234–45.

Hall, S., and L. Appleyard. 2012. Financial business education and the remaking of gendered investment banking subjects in the (post-crisis) City of London. *Journal of Cultural Economy* 5 (4):457–72.

Harvey, D. 1982. *The Limits to Capital*. Oxford: Blackwell.

Harvey, D. 2001. *Spaces of Capital: Towards a Critical Geography*. Edinburgh: Edinburgh University Press.

Harvey, D. 2011. Roepke lecture in economic geography: Crises, geographic disruptions and the uneven development of political responses. *Economic Geography* 87 (1):1–22.

Ho, K. 2009. *Liquidated: An Ethnography of Wall Street*. Durham, NC: Duke University Press.

Ingham, G. 2004. *The Nature of Money*. Cambridge, UK: Polity Press.

Johnson, J. 1966. The money = blood metaphor, 1300–1800. *The Journal of Finance* 21 (1):119–22.

Langley, P. 2004. (Re)politicising global financial governance: What's 'new' about the 'new international financial architecture'? *Global Networks: A Journal of Transnational Affairs* 4 (1):69–88.

Langley, P. 2008a. Sub-prime mortgage lending: A cultural economy. *Economy and Societ,* 37 (4):469–94.

Langley, P. 2008b. *The Everyday Life of Global Finance: Saving and Borrowing in Anglo-America*. Oxford: Oxford University Press.

Langley, P. 2013. Anticipating uncertainty, reviving risk? On the stress testing of finance in crisis. *Economy and Society* 42 (1):51–73.

Langley, P. 2015. *Liquidity Lost: The Governance of the Global Financial Crisis*. Oxford: Oxford University Press.

Langley, P. 2016. Crowdfunding in the United Kingdom: a cultural economy. *Economic Geography* 92 (3):301–21.

Langley, P., and A. Leyshon.2012. Financial subjects: an introduction. *Journal of Cultural Economy* 5 (4):369–73.

Lazzarato, M. 2012. *The Making of the Indebted Man*. Los Angeles, CA: Semiotext(e).

Lazzarato, M. 2015. *Governing by Debt* Los Angeles, CA: Semiotext(e).

Lee, R. 2011. Spaces of hegemony? Circuits of value, finance capital and places of economic knowledge. In *The Sage Handbook of Geographical Knowledge*, eds J.A. Agnew and D.N. Livingstone, 185–201. London: Sage.

MacKenzie, D. 2009. *Material Markets: How Economic Agents are Constructed*. Oxford: Oxford University Press.

MacKenzie, D., D. Beunza, Y. Millo and J.P. Pardo-Guerra. 2012. Drilling through the Allegheny Mountains: Liquidity, materiality and high-frequency trading. *Journal of Cultural Economy* 5 (3):279–96.

Mann, G. 2010a. Value after Lehman. *Historical Materialism* 18 (4):172–88.

Mann, G. 2010b. Hobbes' redoubt? Toward a geography of monetary policy. *Progress in Human Geography* 34 (5):601–25.

Mayhew, A. 2011. Money as electricity. *Journal of Cultural Economy* 4 (3):245–53.

McDowell, L. 1997. *Capital Culture: Gender at Work in the City*. Oxford: Blackwell.

Panitch, L., and S. Gindin. 2013. *The Making of Global Capitalism: The Political Economy of American Empire*. London: Verso.

Pasanek, B., and S. Polillo. 2011. After the crash, beyond liquidity. *Journal of Cultural Economy* 4 (3):231–8.

Patterson, S. 2011. *The Quants: How a Small Band of Maths Wizards Took Over Wall Street and Nearly Destroyed It*. New York: Random House.

Peckham, R. 2013. Economies of contagion: financial crisis and pandemic. *Economy and Society* 42 (2):226–48.

Pettifor, A. 2006. *The Coming First-World Debt Crisis*. Basingstoke: Palgrave Macmillan.

Politi, J. 2008. Senators promise action this week. *Financial Times*, 1 October, p. 8.

Simmel, G. 2004. *The Philosophy of Money*, 3rd edn. Ed. D. Frisby, trans. T. Bottomore and D. Frisby. London: Routledge.

Stäheli, U. 2013. *Spectacular Speculation: Thrills, the Economy, and Popular Discourse*. Stanford, CA: Stanford University Press.

Taylor, M.C. 2004. *Confidence Games: Money and Markets in a World Without Redemption*. Chicago, IL: University of Chicago Press.

Thrift, N. 1994. On the social and cultural determinants of international financial centres: the case of the City of London. In *Money, Power and Space*, eds S. Corbridge, R. Martin and N. Thrift, 327–55. London: Blackwell.

Turner, A. 2009. *The Turner Review: A Regulatory Response to the Global Financial Crisis*. March, Financial Services Authority, London.

Watson, M. 2007. *The Political Economy of International Capital Mobility*. Basingstoke: Palgrave Macmilllan.

Zelizer, V. 1994. *The Social Meaning of Money: Pin Money, Paychecks, Poor Relief and Other Currency*. New York: Basic Books.

4

Making Financial Instability Visible in Space as Well as Time

Towards a More Keynesian Geography

Gary A. Dymski*

Introduction

This chapter explores why the growing body of geographic research on finance, stimulated to new heights by system-shaking financial crises, has neglected financial instability. Bringing financial instability into focus as an element of geographic financial discourse means making space therein for a consideration of the impact of real time and uncertainty. These elements, at the core of Keynes' ideas, are needed if money and credit are to play an essential analytical role in geographical research on finance, and as such they provide the basis for conceptualizing financial instability. So bringing financial instability into spatial analysis requires a more Keynesian geography.

This chapter's theme echoes that in an essay written more than two decades ago by Corbridge and Thrift (1994). These authors argued that money, despite its being at the centre of recent global crises, had been neglected in geography. Harvey's pioneering work in the early 1980s had showed how money and credit relations are central to capitalist – and especially urban – dynamics. But Corbridge and Thrift objected to

* The author thanks the co-editors and Ron Martin for immensely helpful comments on earlier versions of the ideas presented here. Remaining errors are the sole responsibility of the author.

Harvey's unwavering hypothesis: that while financial phenomena accompany capitalist crises, these crises are moved by deeper causes – and thus financial phenomena are of interest only insofar as they are linked to capitalist crisis. These scholars' volume helped launch financial geography as a subfield within the discipline; a range of topics, ranging from financial exclusion to pension-fund investment, was explored.[1]

This enhanced academic inquiry accompanied the growing prominence of financial activity within capitalist economies – a phenomenon denoted as 'financialization'. Harvey (2003) used the term to denote the increasing prominence of financial speculation and motivations within the dynamics of US capitalism. This usage corresponded with Krippner's (2005, p. 173) definition: 'the rise of finance as a macro-level phenomenon shaping patterns of accumulation in the US economy'. However, Epstein created a broader context for this term when he introduced his iconic definition wherein 'financialization means the increasing role of financial motives, financial markets, financial actors and financial institutions in the operation of the domestic and international economies' (Epstein 2006, p. 3). This broad definition, contrasting with the Harvey/Krippner linkage of financialization to a stage of capitalist development, gave rise to a literature so diverse in analytical intention and content that Christophers (2015a) has complained that the term 'financialization' has come to mean everything and nothing.

The onset of the subprime crisis in 2007 and of the European financial crisis in 2010 spurred much new work in geography on how financial systems operate in and across space. Geographers have helped to document the fundamentally spatial character of the subprime crisis (Aalbers 2009; French, Leyshon and Thrift 2009; Martin 2011). This work has emphasized how the trajectory of this crisis was shaped by the social and market power that has accrued to the financial industry and is concentrated in global financial centres (Wójcik 2013). Geographers are also exploring the spatial dimensions of the Euro-area crisis (Christopherson, Clark and Whiteman 2015).

Beyond documenting their spatiality, geographers have begun asking how financial crises fit into the play of economic, political and social forces across space – at the national, regional and local levels. In different ways, investigators have attempted to 'follow the money' (Christophers (2011a, 2011b) and to 'follow the credit' (Gilbert 2011). As with financialization, geographers have taken very different paths to comprehending financial crisis. Three different directions in post-2007 geographic analyses of financial crises are identified in the section 'Geographers on Money and Credit' further below. What is absent in all three approaches is any attention to financial instability or to Hyman Minsky, the economist who originated this concept. This lacuna is documented in Dymski (2017), who shows that the terms 'financial instability' and 'Minsky' – the originator of the 'financial instability hypothesis' – appear only three times and twice, respectively, as Web of Science keywords for articles published in ten leading geography and regional studies journals in the 2006–14 period.[2]

Financial instability can be defined, in Minsky's words, as follows: 'Instability of finance markets – the periodic crunches, squeezes, and debacles ... is a normal functioning internally generated result of the behaviour of a capitalist economy' (1978, p. 2).[3] Two key points here are that these periods of crisis are endogenous to normal capitalist dynamics, and that it is the behaviour of private firms that drives this process: the financial system in a capitalist economy is necessarily on an unstable trajectory, so that loss, dislocation and failure are normal accompaniments of the accumulation process. Minsky sees this approach – his 'financial instability hypothesis' – as derived from Keynes' insights, as having an inherently macroeconomic dimension, and as opposed to equilibrium-based approaches to macroeconomic and financial equilibrium. We retain this extended meaning in describing financial instability herein.

The absence from geography of financial instability – which, as noted, rests on real time and Keynesian uncertainty – has many important consequences. The absence of attention to real time and uncertainty in conceptualizing finance leaves geographers without an analytical bridge to core Keynesian and Minskyian ideas, and thus to the work of the primarily heterodox economists who reject equilibrium and insist that financial instability emerges endogenously in capitalist dynamics. It also constitutes a barrier to geographers being more centrally engaged in policy debate about the global financial crisis, frustrating the intent to engage so clearly expressed by Pryke (2011): 'money, finance, power and space are all alive and well, as it were, albeit reconfigured and reconfiguring ... clearly central to how the contemporary world works and why it doesn't ... now ... is the time for geographers of finance to make significant contributions to research and policy agendas.'

Geographers who follow Harvey's Marxian approach, inasmuch as they focus on power but ignore endogenous instability, tend to discount financial policy discourse: financial crises represent one manifestation of an ever-deepening crisis of capitalist relations *tout entier*. Economists who follow Minsky, by contrast, see capitalism as inherently fragile and subject to recurrent financial crises that do not imply its imminent collapse; economic policy to address financial instability matters. But in the absence of a bridge to these economists, geographers engaged in economic policy analysis have relied on ideas generated in non-Keynesian equilibrium-centric models, which equate policy analysis with choice among contrasting microeconomic policies. Consider geographers' recent interventions into how to stimulate economic growth: on one side are proponents of the New Economic Geography advocating support for industrial clusters, on the other are calls for more localized economies and more balanced regional growth (Martin 2015).[4] Geographic analysis consequently has no conceptual leverage with which to critique financial instability or austerity macro policy.

In sum, breaking geographers' silence on policy alternatives regarding financial instability and macroeconomic policy will require, as a precondition, bringing Keynesian ideas into geographic work on money and finance. The section below, 'The Forgotten "Lessons" of the Financial Crisis' reviews the policy 'lessons' of the financial crisis, which have subsequently been ignored. The next section 'How the Core Concerns of Monetary Economics Disappeared from Mainstream Economics' locates a root cause of this tepid policy response: the disappearance from mainstream economics, in the post-war period, of essential roles for money and credit in theories of economic growth. This coincides with the privileging of general-equilibrium methods in economic analysis. 'Why and How Money Matters: Towards a More Keynesian Geography' shows that reviving an essential role for money and credit requires an analytical framework in which power, real time and uncertainty matter. 'Geographers on Money and Credit: Seeing with Political and Cultural Lenses' argues that geographers' analysis of money and credit issues, while often prioritizing power, typically overlook real time and uncertainty, and thus consign money, credit, and macroeconomic policy to marginal analytical roles. The final section concludes the chapter.

The Forgotten 'Lessons' of the Financial Crisis

The subprime crisis of 2008 exposed the weaknesses of the contemporary structures of money and finance in the most brutal terms: many nation states were forced to provide public subsidies or buyouts for insolvent banks and financial firms so large and so complexly interconnected with global financial markets that their failure could not be contemplated.

To understand how such sophisticated and technologically-advanced markets could so profoundly fail, policymakers and market players alike turned initially to ideas about financial instability developed in the 1970s by maverick Keynesian economist Hyman Minsky. Simultaneously, the model that had dominated mainstream economic thinking since the early 1980s – which viewed the aggregate economy as a continuously growing equilibrium maintained by rational market actors with rational expectations – came under fierce attack, both by fellow economists (Krugman 2009) and by others (Fox 2009).

Minsky (1986) had argued that when financial breakdowns occur, *big bank* and *big government* intervention is essential: central bank 'lender-of-last-resort' liquidity provision to stop speculative runs on financial markets; and then counter-cyclical spending to sustain aggregate demand while damaged private-sector balance sheets recover. Minsky argued for these drastic actions even while the regulated financial system of the post-war years remained in place. In an article published just prior to his death, Minsky (1996) foresaw that what he termed "money market capitalism" would generate still further instability. And indeed, the 2007–08 crisis generated a meltdown of the remarkably complex, hyper-leveraged, underregulated financial system that had resulted from a quarter century of deregulation. The US and other affected governments did use both 'big bank' and 'big government' interventions, per Minsky's definitions, to bring the meltdown under control. This crisis, it seemed clear, brought further lessons about the dysfunctionality of finance:

1. Underregulated financial intermediaries, especially outside the formal banking sector, could undermine growth and compromise the non-financial (real) economy;
2. Regulatory provisions for greater transparency and protections against predation of financially fragile economic units, should be put in place immediately;
3. Coordinated governmental action was crucial in preventing local-ized financial losses and financial panics from turning into general-ized depressions;
4. More realistic financial models, such as Minsky's framework, should guide economic policy and displace models based on continuous equilibrium.

Minsky's work had a brief resurgence as the crisis unfolded (Cassidy 2008; Wolf 2008). But none of the lessons based on his ideas stuck. Instead, financial reform efforts were slow and tepid. In the United States, the Dodd–Frank Act became law only in July 2010. Its key provisions were watered down – especially those focused on the protection of consumers' rights, on limiting unsecured risk-taking and proprietary trading by large banks, and on unifying regulatory authority over shadow banks' activities. In Europe, a reorganized bank regulatory structure emphasized capital adequacy. In the US, the UK and Europe, banks retained substantial free-dom of action, regulatory authority remained fractured or incomplete, consumer rights were subordinated to bank solvency, and only minimal cross-ownership and skin-in-the-game provisions were implemented.

Far from being punished for sponsoring unsound instruments, many banks and non-banks, especially the largest, offloaded huge amounts of their fixed-income instruments onto central bank balance sheets via 'quantitative easing' programmes launched initially in 2009. Despite these measures, banks lent very cautiously: mortgage lending depended on government support, and hedge funds accounted for most commercial and industrial lending growth.

Finally, the resistance to economic orthodoxy did not take deep root. Minsky's work was forgotten as easily as it had briefly been remembered. Within a matter of months, mainstream economists were creating new generations of models with incomplete information, transaction cost and asymmetric information features capable of explaining aspects of the great financial crisis.

How the Core Concerns of Monetary Economics Disappeared from Mainstream Economics

This failure to heed the lessons of the recent crises about financial dysfunctionality was prefigured by the disinterest of mainstream economic theory in the role of money and credit in economic processes. There is some irony here. For while virtually every introductory economics text since Samuelson (1948, pp. 57–8) has contained a section on the 'functions of money' – to serve as a medium of exchange, unit of account, and store of value[5] – those at the cutting edge jettisoned the functions of money from economic theory long ago.

After World War II and the creation of national income accounts, textbooks incorporated macroeconomics, addressing the functioning of the economy as a system. Incorporating money and banking was crucial in explaining how the system as a whole worked. The first economics textbook published in the post-war United States, Lorie Tarshis' *Elements of Economics* (1947), provides an institutional account of how commercial banks create money in the course of their loan making, in a chapter entitled 'Money and Commercial Banks'. Samuelson discusses the functions of money in his chapter 3, 'Functioning of a "mixed" economic system', in a section on 'Exchange, division of labor, and money'.[6] The argument implicitly appeals to the structural-functionalist social theory of mid-century anthropology (Malinowski 1926; Radcliffe-Brown 1958) and sociology (Parsons 1966), which in turn built on the work of Durkheim (1893).

But the economics profession's search for scientific status left the functions of money approach, along with its unstated assumption that economic mechanisms operate within social wholes, abandoned on the lonely islands of introductory textbooks. Samuelson himself led the way;

his *Foundations of Economic Analysis* (1947), borrowing heavily from physics (Mirowski 1991), reimagined economic dynamics not as interacting markets, as in Marshall (1920), but instead as trajectories in economic subspaces. The 'Holy Grail' quest for twentieth-century economic theory became the proof of the existence of Walrasian general equilibrium (WGE): the idea that a set of individual economic agents, making decisions solely on the basis of their own preferences and wealth endowments, can make decisions based on a set of projected prices that, when added up, yield demand–supply equilibrium in every market. The actual exchanges can unfold over time, but the idea is that all the decisions about exchange, production and consumption occur simultaneously.

This quest succeeded partially: mathematically, the existence and uniqueness of such an equilibrium could be established, but not its stability (Ingrao and Israel 1990). The conditions required for this equilibrium included timelessness – everything happens at once – spacelessness – there are zero transaction costs – and complete information – everything pertinent to economic processes is known to all agents in the economy, and every agent knows what any other agent knows. In a world of instantaneous exchange, in which it is costless to use markets, money and credit are extraneous. Any commodity can play the role of 'numeraire' (unit of account). No credit market is needed because loans taken out now against future income are self-liquidating.

Through the years of the Fordist era, macroeconomics was an applied policy area, only vaguely linked to considerations of choice and equilibrium. For pragmatists like Samuelson, it was a matter of leaving well enough alone.[7] President Richard Nixon's famous 1971 declaration that he was 'a Keynesian in economics' came as no shock; but the dollar-based Bretton Woods system came under such pressure that its dismantling began that same year. The idea that government policy could fine-tune economic growth by exploiting the Phillips curve – a stable tradeoff between inflation and unemployment – came under increasing pressure. James Tobin defended the Phillips curve, in his 1971 presidential address to the American Economic Association, by setting out a 'theory of stochastic macro-equilibrium' which acknowledged the difficult policy context:

> stochastic, because random intersectoral shocks keep individual labor markets in diverse states of disequilibrium; macroequilibrium, because the perpetual flux of particular markets produces fairly definite aggregate outcomes of unemployment and wages. (1972, p. 9)

This rationalization failed to stop attacks on Keynesian macropolicy by Robert Lucas and other members of the emerging New Classical approach to macroeconomics. In their view, Keynesian macroeconomics

had failed because it lacked microfoundations based on 'rational expectations'.[8] They argued that since rational economic agents were already
in equilibrium whenever the government initiated any policy scenario,
those agents would react against that policy shift and undo it. In their
view, government was either powerless or the enemy of equilibrium.[9]

Henceforth, both micro and macro theory would be judged on the
basis of whether they incorporated microfoundations. These involved
establishing that individual economic agents inserted into whatever context the theorist chose would be depicted as taking only actions that
maximized their self-interest. If the world involved timeless, spaceless,
complete-information markets, of course, no policy would ever be
needed. Of course, the world deviated in numerous ways from that
desideratum; and economists proceeded to establish new bases for
economic policy interventions – everything from monetary policy to
industrial policy – by arguing that one or another deviation from the
conditions required for WGE required it. Indeed, this was how one's
conformity with the scientific pretensions of the mainstream was demonstrated. Macroeconomics became, as Farmer (1993) put it, 'an applied
branch of general equilibrium theory'.

So roles for money and credit could be justified, as economics mod-
eling parlance has it, using well-chosen deviations from WGE: money, by
introducing the assumption that transactions are costly, due to factors
such as spatial or temporal distance between economic agents; credit, via
an assumption that everything does not happen at once, so there are mis-
matches in time between spending and income. But given mainstream
economics' insistence on rigour, and the standard of rigour being the
Walrasian general equilibrium, any model incorporating assumptions
that provided a role for money or credit should deviate as little as possible from that reference model. The modeller's challenge was to identify
the minimal distance from the WGE required to demonstrate a certain
outcome as necessary. Such outcomes are then termed 'second best'.
These outcomes could include social institutions. For example, banks'
existence can be explained through two deviations from WGE. First,
assume there is asymmetric information about borrowers' willingness to
repay a loan (borrowers know their own intention of cheating, or not,
and those who will cheat have no incentive to tell the truth); this requires
that lenders monitor borrowers' performance. Second, assume there are
economies of scale in monitoring; so one agent (the 'bank') pools savings
from prospective lenders and accomplishes the required monitoring at a
socially minimal cost.[10] Introducing more 'realism' – further deviations
from the conditions required for WGE – only compromised understanding.
Adding a stochastic dimension to any analysis changes nothing, unless
the agents in the model are risk-averse.[11]

A key insight here is that this approach can explain social outcomes, and even institutions and voting patterns, by working backward from individuals' rational choices under various constraints. The triumph of this mainstream vision is that it can explain society without any need for a theory of society; it relies only on the 'primitives' – individuals' (pre-social) preferences, their wealth endowments and the state of technology.

Seeing the 2008 financial crisis

The argument set out in this section provides a sufficient basis for understanding how economists working from mainstream premises viewed the crisis of 2007–08. By the time the crisis occurred, the agreed macro-modelling framework – the dynamic stochastic general equilibrium model – envisioned macroeconomic growth paths as representing the optimizing choices over consumption, investment and savings options of representative-agent-based economies. The problem of these agents was to coordinate their own behaviour over time – earn now and consume later, or vice versa? An overall balanced-budget requirement obtained for each agent (you can only spend what you earn). Government could rearrange expenditures, but only by removing expenditure opportunities (taxing agents) elsewhere in the time sequence of agent lives. This was the microfounded general equilibrium approach that Farmer had anticipated. In the principal advanced textbook on macroeconomics in use – Woodford (2003) – money is treated only as an abstract symbol of purchasing power; store-of-value considerations are unnecessary because agents are ultimately coordinating only with themselves.

So even an event as profound as the subprime crisis does not signify for economists that the equilibrium model itself is broken (Mirowski 2013). The problems lie elsewhere. Mian and Sufi (2014) note that most mainstream economists' accounts of the subprime crisis see it in one of three ways: first, the result of an exogenous shock; second, the fact that people in the economy had excessively optimistic assessments of future housing prices ('animal spirits'), which were deflated along with housing prices to cause the crisis, and; third, bad monetary policy.[12] These authors argue that excessive and poorly structured household debt is the real culprit. Contractual design improvements and the discouragement of excessive household indebtedness will avoid any repeat of the problem. Note that this is a shift to the microlevel of economic analysis. These authors do devote a chapter to macroeconomic policy. In this chapter, they observe that monetary policy can no longer reliably prevent episodes of overlending from turning into crises; and fiscal policy is a 'clumsy alternative' representing 'an attempt to replicate debt restructuring' (pp. 163–4): that is, its intended effect is to transfer spendable income from creditors to debtors so as to bolster aggregate demand in the shorter run. There is no engagement with macroeconomic theory here; but the comments made conform to the prevailing model summarized above – expenditures can be rearranged temporally or spatially, but only 'clumsily'. And in the long run, what was borrowed has to be repaid, whether by households or by governments.

Apart from Mian and Sufi, numerous other mainstream explanations of the subprime crisis have been proposed. Some attack government efforts to channel credit for social welfare purposes (Calomiris 2014). Many others propose yet other mechanism-design solutions, based on flaws that stem from inadequate market outlets and/or missing information in markets (or both, as in the explanation of Shiller 2008), inadequate market depth, transactions costs or other incentive problems. It can be noted that this rush to explain financial crises either as due to bad monetary policy, agent nerves or to design and incentive problems in credit markets also occurred in the wake of all the other major neoliberal-era financial crackups: the 1982 Latin American debt crisis, the 1981–89 US savings-and-loan debacle, the 1997 East Asian financial crisis.[13] The edifice of the macroeconomic model itself – its articulation of macroeconomic coordination as a large-scale application of the general-equilibrium problem of coordinating investment, consumption and savings decisions across time – is never directly challenged, except in the heat of the moment (and then only in the subprime crisis). Once consensus emerges on what specific design-mechanism flaws caused the problem, normal science can go on.[14]

Why and How Money Matters: Towards a More Keynesian Geography

So in the mainstream approach, money and credit affect economic outcomes only when there are malfunctions in one or more of the specific design-mechanisms by which they are delivered. Isn't this good enough? Shouldn't introducing money and credit into an economic model simply mean identifying a qualifying deviation from Walrasian general equilibrium, and then getting on with it? To answer 'no' to this question, it is necessary to ground economic analysis in a conceptual field other than general equilibrium. This means returning to the starting point that political economy shared with sociology and anthropology before the physics-inspired 'marginalist revolution' (Mirowski 1984): social relations – society itself – as the central object of study.[15]

Ingham (2004) has observed that sidestepping this very conceptual embedding is what permits mainstream (or, in his terminology, orthodox) economists to relegate money to an epiphenomenal status: 'the questions of what money did and its cultural significance for modern society were addressed [by sociology and history]', but its nature was not seen as problematic. Indeed, 'it is held that money is a "neutral veil" over the workings of the "real" economy' (7–8). But, he argues, this leads to a huge category error:

> The questions of how money was produced and how it was able to perform its functions were rarely posed. It was generally accepted that the ontology of money was adequately dealt with by the venerable theory in which money's functions were deduced from its status as a commodity... this entailed a serious logical category error. Such functions cannot be established in this manner; rather, they are institutional facts that can only be assigned in the construction of a social reality. (Ingham 2004, p. 197)

In consequence, 'orthodox economics has failed to *specify* the nature of money' (8). And this, in turn, has led the other social sciences to a similarly shallow treatment:

> Yet the other social and historical sciences have fared no better in the analysis of money. As a direct consequence of the division of intellectual labour in the social sciences ... they have been unable to provide a more satisfactory account. In the mistaken belief that it is essentially an 'economic' phenomenon, the other social sciences have abnegated all responsibility for the study of money, by either simply ignoring it or uncritically accepting orthodox economic analysis. (9)

The 1994 *Handbook of Economic Sociology* provides a dramatic, if somewhat indirect, illustration of mainstream economics' hegemonic influence over the way that other social sciences see money and credit processes. Prominently positioned in this volume is a chapter by Oliver Williamson on transactions-cost economics, on which the increasing-returns models that have shaped the new economic geography are based. Williamson differentiates among topics which are or are not suitable for economic analysis based on whether they are conformable with rational choice approaches. Ingham (2004, p. 199) summarizes the perspective that informs this publication as follows, 'Time, history, and culture are analytically unimportant; they are the "contextual tosh" from which micro-economic analysis can extract the "rational core".'

Thus if money and credit are to be fundamental, and not epiphenomenal, in economic relations, we must identify more specifically how money and credit emerge and what they signify in the social reality within which the economy is embedded. This can be done by engaging with some of the deep controversies in monetary theory. One defining divide emerged in the mid-nineteenth-century British controversy between the 'currency school' and 'banking school' (Skaggs 1999). In the former view, money is created and controlled by government, and this extends to the volume of credit; so any problems that arise from this quarter are due to ill-advised government policies. In the latter view, credit is created by banks endogenously, in volumes determined by their social and economic linkages with current and potential borrowers. Credit then can concatenate independently of the will of government, and its creation is institutionally contingent.

So the 'currency school' repeats the mainstream duality between market and government; the latter brings money and credit into the social context. Opting for the 'banking school' leads, in part, to a map of power: who has the power to create money, and who can compel the repayment of credit contracts, and decide the consequences of non-payment? Ingham too emphasizes the power dimension of money and credit, writing that 'it is not simply a question of the possession and/or control of *quantities* of money – the power of wealth. ... the actual process of the *production* of money in its different forms is inherently a source of power' (Ingham 1999, p. 4).

This power dimension of financial relations is acknowledged by geographers engaged in analysing financial crises, as discussed in the following section.

What is largely missing in geographers' discourse, however, is a second dimension along which 'currency' and 'banking' schools differ: attention to the informational basis of the world in which we live. Is it a world of repeated events and cycles, or of uncertainty? That is, is it more fundamentally a matter of who possesses the needed information, or is it that the needed information doesn't exist at all? Is time 'notional' or 'real' – is a map of all possible outcomes and their probabilities out there, waiting to be discovered; or does this map not exist? If it is the latter, then many consequences follow. In particular, one consequence is that agents cannot make decisions based on reliable information. They form expectations of the future based on a set of conventions that are created socially, and held with variable degrees of confidence. Whilst agents seek financial returns, they also fear loss, and when conventions collapse, run from commitments that lock in future actions. Then they will seek liquidity. The fact that people can run from spending the resources they control means that aggregate demand can fall short of available supply. So there is an aggregate level of economic activity that varies and affects overall economic conditions, as a consequence of the exercise of liquidity preference. These, of course, are fundamental insights of Keynes.

Placed into temporal order, Keynes fits in just after Marshall in economics' temporal sequence from the theory of society to Marshall to general equilibrium. But Keynes' *General Theory* (1936) places him outside any linear progression. Indeed, its two interlocked themes – that economic malaise in capitalism can be traced to shortfalls of aggregate demand and to fundamental uncertainty – violate the very premises of orthodox economic models. But these are, at the same time, precisely the analytical features that make money and credit central to the alternative vision of capitalist economic dynamics that he constructs. Keynes argues that because the future is unknowable, agents – who must still decide whether to undertake longer-term commitments (notably investment) or to defer them – rely on conventional opinion and the market(s) in question, with a variable degree of confidence. One or more adverse outcomes can crush confidence and erode conventions; in that case, agents will minimize spending and maintain flexibility. This is, then, the root of Keynes' theory of money: money is held, of course, to make transactions; but beyond this, it can be held for precautionary purposes – as an expression of agents' liquidity preference.

It is readily apparent that this ('banking school') theoretical approach is rooted in a conception of the economy as an inherently social construct. Aggregate demand can fall short of aggregate supply – an impossibility in a price-coordinated equilibrium – when agents choose to save more, and invest or consume less, then they normally do. Credit – wherein banks or economic units seek to lend money now for return later – has an extra degree of danger attached; beyond the problem of how well the lender knows the borrower's intention (as above) is the question of whether the lender is willing to make a contract whose terms permit borrower repayment only after a non-trivial period of time has passed. Keynes, in effect, imagines a world of real time, not just notional time, wherein events that will come to pass later cannot be forecast with confidence, even using the probability calculus. 'We simply do not know', as he famously wrote (Keynes 1937, p. 214).

Some key elements of economic growth and stagnation in 'real world' economies become visible once the power to create credit or to withhold it, as well as real time and uncertainty, are taken seriously as features of economic environments. The 'lessons' of the financial crisis – so easily forgotten by economists working from orthodox premises – are not paradoxes to be picked apart by identifying the specific factors that may have disturbed financial equilibrium processes; they are instead logical outcomes when seen from the entry point of Keynesian economic theory.[16]

The recognition that credit-creation is a source of power in the economy immediately opens the possibility that social inequality and exclusion may profoundly affect economic outcomes; that is, incorporating what might be termed the 'social construction of creditworthiness' (Dymski 1998) requires accounting for the interaction of social and economic dynamics. This implies that money is not neutral, as Ingham emphasizes; thus, asymmetric distributions of power in creating and being responsible for the repayment of credit can generate wildly different social and economic outcomes. Uncertainty itself has profound consequences: economic equilibrium itself is unattainable; the level of aggregate demand is prone to fall short if it depends on either capitalists' investment or households' consumption; confidence and beliefs are fragile and can crumble, leading in turn to shifts in liquidity preference that can destabilize financial markets and undermine economic growth. Hyman Minsky (1986) put these two elements, credit-power and uncertainty, together in his descriptions of the ways in which financial commitments in times of great confidence can exceed what can realistically be repaid, leading to financial crisis. And in his further reflections on financial instability, Minsky (1996) showed how continuing institutional evolution in finance can lead to more profound and possibly more uncontainable financial crises.

Seen from a 'banking school' perspective, power, real time and uncertainty should logically be included as core elements of an economic model, since these are features of the world in which we live. Including these elements in a model permits a full consideration of the role of money and credit in economic dynamics; excluding them thwarts this purpose. To exclude power and uncertainty in analysing the economic is to describe the surface but not the core of financial processes.

Geographers on Money and Credit: Seeing with Political and Cultural Lenses

Geographers, when analysing money and finance, have often emphasized that power is manifest in political forces and cultural practices; however, they typically overlook real time and uncertainty. This attention to power, but inattention to uncertainty, is evident in work written and published both before and after the subprime crisis of 2007–08, and in work analysing financial crisis itself. We consider these subliteratures in order.

Pre-crisis

Geographers' chronic inattention to Keynesian ideas is evident in their characteristic use of the term 'Keynesian'. The essay *Money, Power, and Space* by Corbridge and Thrift surveying the 1994 volume they co-edited with Martin describes its contents as follows: first, a geopolitics section examining the money and financial aspects of the power relations that underlie the organization or reorganization of production and exchange; second, a section on 'institutional geographies of international finance and national regimes of industrial and financial capitalism', including attention to 'the institutional frameworks in and through which industrial and financial capitals are brought into local (and not so local) contact'; third, a section on the politics of money, 'asymmetries of power, which must vary from place to place'.

Ron Martin mentions financial instability in his chapter in this volume; but he does so primarily in a political context: 'Financial globalization is ...reconfiguring the geographies of money, power, and dependency,[leading to] ... the increased disarray and instability that now characterizes the world financial system ... and it is *within the nation state* that the instabilities of global money appear' (Martin 1994, p. 274).

Corbridge's chapter on inflation, which contrasts Monetarist and Keynesian approaches, openly admires Keynesian economics:

> Keynesian models are commended to human geographers because of their clear attachments of political economy, because of their evident concern for a practicable politics in an uncertain world economy, and because of the extraordinary relevance of Keynes's own work to the 1990s. Keynes never set much store by abstract economic principles – neither *laissez-faire* nor state socialism – and his studied eclecticism (not to mention his voice) is celebrated openly throughout this chapter. (Corbridge 1994, p. 64)

He then explains the absence of attention to Keynesian economics within geography when he observes:

> geography for the past twenty years has been inspired by Marxism above all else. ... One effect of this attachment to Marxism, however, has been a willingness to see critique mainly in destructive terms and with little regard for its constructive dimensions and responsibilities. Insofar as radical geography tends to associate most of the ills of the modern world with 'capitalism', it has sometimes seemed to suggest that the abolition of capitalism will provide a corresponding cure-all. [But] all forms of critique are weakened to the extent that they do not also provide plausible accounts of how human affairs might better be arranged or managed. (Corbridge 1994, p. 87)

Corbridge sees Keynesianism as embodying the 'pragmatism' he is calling for, and concludes, 'Away from the eclectic mainstream of social science, it sometimes seems that a rigorous eclecticism is more often reviled than admired' (88). This said, Corbridge does not elaborate further about the elements of which a Keynesian approach to economic geography would consist.

The *Oxford Handbook of Economic Geography* (Clark, Feldman and Gertler 2000) contains two chapters on 'finance capital' in its section on 'global economic integration': one (Laulajainen 2000) discusses the regulation of international finance, whilst the other (Tickell 2000) examines the links between finance and localities. The key emphasis of both essays is captured by Tickell's concluding observation that 'financial geographies are also geographies of power and are critical in moulding the quality of people's lives' (243). The chapter by Sheppard (2000) discusses time in ways that move toward the Keynesian conception of real time, but without any mention of money. Peck (2000) also recognizes the importance of 'time and space' in his chapter; and, like Corbridge, he signals the absence of a conceptual centre in geographers' work on financial questions – but without posing a solution:

> ever since economic geography abandoned neoclassical theory in the 1970s, the subdiscipline has been preoccupied ... with the movements of the 'real economy' over time and space. ... [But b]ecause ... institutional relations and outcomes are contingent, they cannot be exhaustively theorized in advance and their effects cannot be predicted with certainty.
>
> Correspondingly, economic geographers are rarely prone to make sweeping statements about the anticipated effects of institutions or policy measures. They tend, by inclination, to be suspicious of such predictive, 'big-picture' generalization. ... [but beyond] spoiling the parties started in other disciplines ... there is a need for economic geography to codify what it *does* know about economic institutions (and how it found out). (75)

Post-Crisis

In the 2011 *SAGE Handbook of Economic Geography* (Leyshon, Lee and McDowell 2011), the term 'Keynesian' is used solely to describe a welfare state regime of capitalism; Keynes' foundational work on the nature of money and financial markets does not appear. This volume contains numerous mentions of 'finance', primarily in the context of the role of financial centres in establishing levers of control over capitalist dynamics in the neoliberal age. One chapter (Pryke 2011) focuses on money and finance. Pryke describes how finance came to play a 'headline act' role in geography. The extensive financial crises of the 1980s and 1990s, he notes, led leading geographic thinkers – singling out Harvey, Thrift and Leyshon – to turn their attention to the 'processes generating remarkably uneven life chances' in 'economies run on and for the needs of high octane finance' (Pryke 2011, p. 287). He describes the way in which changes in global and national financial regulation, combined with technological advances, altered financial architectures and created 'real financial geographies' that involve 'an intricate and paradoxical interplay between flows and territories'. Financial centres such as the City of London remain crucially important in these evolving architectures, even while new space is continually opened up by innovating financial firms exploiting new informational and digital technologies.

Leyshon's chapter argues that economic geography has been caught 'in a pincer movement, between two more theoretically rigorous movements: Marxism informed political economy on the one hand, and the more theoretically catholic body of work concerned with economic agglomerations on the other' (Leyshon 2011, p. 389). His essay suggests breaking out of these pincers by following the lead of those social-science disciplines which have recognized that 'the way in which what we understand as "the economic" [is] constituted through a range of cultural, social and political practices' including 'the possibility of agency, and the ability of powerful actors to intervene within the unfolding of the economy, albeit within a broader context' (ibid., p. 394).

Geographic reflections on the recent financial crises

Geographers have undertaken three distinct – if by no means mutually exclusive – approaches to the twenty-first-century subprime and European crises. As with the work just considered, considerations of power are emphasized; but considerations of real time and uncertainty are largely absent.

The first approach, embodied by Harvey (2014), maintains the view that the financial crises of recent years are manifestations of the deeper contradictions of capital and of capitalism. So financial crises are important insofar as they are sufficiently severe to threaten the reproduction of capitalist accumulation. Capital, for its part, has an imperative – as per the title of Harvey's 2003 book – and state actions, ranging from the encouragement of homeownership prior to 2007 to the bail-out of too-big-to-fail banks, are best understood as means of furthering this imperative.

A second view sees financial crises in a political economic light. Christophers, for example, has followed up his recent examination (2013) of the financial industry's growth across national borders with an essay (2015b) that puts the financial crises to which this industry is so prone in recent years into a more systematic perspective. He argues that financial crises are best viewed through a political economy lens, that is, by focusing on the 'structural and systemic dynamics of capitalism as a mode of socio-economic production and reproduction'; and this leads to a definition of financial crisis as a 'significant diversion from a more or less established historical pattern of financial dynamics and structures' (ibid., p. 463). He then proceeds to show that these diversions have emerged in the past several decades from governmental and regulatory efforts to manage the contradictory challenges posed by the growth of the financial sector, incessant deregulation and technological change. This conclusion closely resembles that reached by sociologist Krippner in her recent (2011) book, wherein financialization of the economy was not a deliberate goal of policymakers; instead, 'the state's attempts to extricate itself from the crisis conditions of the late 1960s and 1970s sowed the seeds of the turn to finance in the US economy' (p. 3).

A third view sees financial crises as emerging from the tenuous and possibly unintended interactions of forces. For example, Langley (2015) uses the cultural economics lens refined in the social studies of finance literature to show how state interventions in the unfolding financial and subsequent fiscal crises are best understood as a fragmented set of interlinked but uncoordinated adjustments – not as a unified optimal response to unfolding events. The writings of Mann can also be placed in this category; one recent example is his investigation (2010a) of the links between monetary policy, financial globalization and crisis. This third view does not deny the significance of capitalist dynamics or state policy actions; what it denies is that any master logic – including the playing-out of a contradictory labour/capital dialectic – can be detected as guiding events. To the contrary, it works from the premise that capitalist dynamics may be riven by complexity and indeterminacy so profound as to obscure any capitalist imperative or undercut any state–capitalist alliance. In this context, a 'cultural' economics would explore how people placed in uncertain positions adapt to circumstance. But what is this circumstance? If the capitalist imperative ran aground, and the established financial pattern shifted in 2007, was this steered by any of the historical agents? Or was it a situation wherein firms and individuals and state policymakers were reacting to their own reactions amidst a dynamic beyond any control or predictability? Could it not be said that such dynamics manifest uncertainty which can (and did) result in financial instability – the loss of control over events by all the parties involved in those events, no matter their intentions or their social power?

Of the geographers whose post-crisis work was mentioned above, Mann stands out as being closest to seeing that uncertainty can not only permeate cultural dynamics but can be a root cause of the endogenous financial instability which Minsky described on the basis of his reading of Keynes. Indeed, Mann (2015) has recently explored Keynes' ideas about unemployment and liquidity in some depth, in an analysis that acknowledges the centrality of fundamental uncertainty to Keynes' vision. He then comes close to expressing the centrality of financial instability as an endogenous component of capitalist dynamics in two post-crisis essays. In Mann (2010b) he writes, 'despite massive "devaluations", the value-relation is not itself in crisis'. Then, in a 2013 essay, Mann observes (p. 198):

> there are immensely powerful geographic and social forces that animate modern money that make it capitalist in its very being, but these forces operate at an almost ephemeral scale, and at an institutional temporality, that blur the lines between the general and the particular, between the macro and micro scales. And it does its work unevenly, unpredictably sometimes. But, at least in capitalism, it is never anything less than enormously important, and its very structure has seemed to put working people on the losing end (again, to say nothing of how little they have of it).

This passage is close in spirit and in substance to the essay that Minsky composed in 1996 just before his death. In acknowledging the centrality of both uncertainty and power in financial dynamics, Mann contributes to building the missing bridge described above.

Conclusion

A club of mainstream economists largely dominates policy discussions, using formal models and econometric techniques embodying a coded language with invisible but real rules about what can be proposed and about how this can be said.[17] These economists do not evidently want or need assistance; their models are chiselled to a deceptive simplicity not only to invite add-ons from earnest amateurs, but also to demonstrate to keepers of the equilibrium faith that they have identified the simplest possible reasons why deviations from optimal equilibria may have occurred. Thus, debate unfolds within predefined limits, always focused on fixes for malfunctioning market incentives or poorly functioning regulatory designs.

But this way of approaching financial crises and their causes is itself one of the causes of the financial crisis. If economic models that enter into general circulation can only work with phenomena that can be specified in equilibrium-based models, then disequilibrium phenomena will go unexamined. For when equilibrium becomes the focus of economic theorizing, the roles for money and credit in economic decision-making, processes and outcomes are necessarily trivialized. That is, the models governing policy discourse in mainstream economics have uniformly restricted ideas about the roles of money and credit in economic processes. In this approach, any event that comes into policy focus represents a deviation from an equilibrium trajectory; so finding a policy solution involves understanding what mechanisms induced the deviation and correcting them. Economists have an entire catalogue of possible mechanism problems on which to draw, based on asymmetric information, incentive incompatibilities, transaction costs, time-inconsistency and so on. Fundamental theorems in economic theory dictate that it is always possible to generate a set of micro-mechanisms consistent with optimizing behaviour that can explain a given result.[18]

So long as the policy terrain is restricted to the application of these methods, then identifying policy fixes will remain the domain of mainstream economists. The relationship of geography to financial processes in this case is equivalent to that of regional studies to industrial agglomeration: the only way for practitioners in this field to contribute to the core theory is to set aside the home-discipline's analytical tools and to adopt those of economic theorists. Space is simply a vehicle for conducting signals emitted by economic agents engaged in optimizing behaviour. The challenges posed by the passage of time can be tamed by applying probability theory to repeated experience.

Breaking the monopoly of equilibrium-based economic theory over discussion about financial processes and financial policy requires a different pre-analytical vision of the challenges of organizing economic exchange and reproduction, wherein time and space are understood as untamed entities whose extent and enormity can frustrate rational planning and undo fragile plans (or plans suddenly revealed to be fragile). Geographers have already contributed centrally to understanding economic processes distributed across time and unfolding in space. Indeed, in a powerful demonstration of the power of acute spatial analysis, well before the 2007 financial crisis that surprised mainstream economists, the geographers Aalbers (2005) and Wyly et al. (2006) explored how subprime lending and credit market discrimination constituted geographic manifestations of unequal social power. Geographers' writing post-crisis has emphasized elements erased from equilibrium-based economic models: power relations and the cultural basis of much economic behaviour. This essay has suggested that geographers' insights will be especially valuable, both for theoretical understanding and for policy insights, if geographers become more Keynesian in their analyses of financial dynamics. And it is not simply a matter of recruiting geography to the ranks of a Keynesian consensus already adequate to the current historical moment. Because of their grasp of the interpenetration of social and economic dynamics over space, and of the multidimensional manifestations of power in these dynamics – which Keynesian economists themselves often overlook – geographers are uniquely positioned to generate new and important insights into the causes and policy implications of financial instability.

Notes

1 For an overview of topics covered in the emerging field of financial geography, see Dymski (2017).

2 The ten journals were *Environment and Planning A, Environment and Planning D,* the *Cambridge Journal of Regions, Economy, and Society, Regional Studies, Regional Science and Urban Economics, Economic Geography, Journal of Economic Geography, International Journal of Urban and Regional Research, Geoforum, Area, Antipode, Progress in Human Geography, Annals of the American Association of Geographers,* and *Transactions of the Institute of British Geographers.* By contrast, the terms 'financial instability' and 'Minsky' each appear forty-eight times as Web of Science keywords in four heterodox economics journals in this same time period.

3 Minsky goes on to elaborate, in an italicized phrase, '*a capitalist economy with sophisticated financial institutions is capable of a number of modes of behavior and the mode that actually rules at any time depends upon institutional relations, the structure of financial linkages, and the history of the economy*'.

4 According to Fujita and Krugman (2004), who co-founded the field, 'the defining issue of the new economic geography is how to explain the formation of a large variety of economic agglomeration (or concentration) in geographical space'.

5 Marshall's iconic *Principles of Economics* (1920) does not mention the functions of money; but the author observes, 'Money is a means toward ends, and if the ends are noble, the desire for the means is not ignoble' (p. 22).

6 Marx's *Capital* (1906) gives money a similar pride of place, in chapter 3 'Money, or the circulation of commodities'. Jevons (1875) describes four functions of money, adding that it is a standard of deferred payments.

7 Davidson (2015) recounts how Samuelson's adherence to Keynesianism involved a less deep intellectual commitment on his part than to models embodying Walrasian rigour.

8 Tobin, himself committed to mainstream economics (and thus unable to free himself from the Keynesian/Classical contradiction) saw the writing on the wall. In the passage immediately after the quote included above, he wrote, 'It is an essential feature of the theory that economy-wide relations among employment, wages, and prices are aggregations of diverse outcomes in heterogeneous markets. The myth of macroeconomics is that relations among aggregates are enlarged analogues of relations among corresponding variables for individual households, firms, industries, markets. The myth is a harmless and useful simplification in many contexts, but sometimes it misses the essence of the phenomenon.'

9 The fact that this equilibrium was unattainable, since these agents' macroeconomic environment had been created by a prior history of policymaking – that is, that there was no 'state of nature' – was ignored. In equilibrium theory, the state of nature to which agents attempt to return is simply assumed to exist.

10 See Diamond (1984), one of the most influential theoretical models explaining why banks exist.

11 The logic is this: if all agents know (or can discover) the probability distribution that governs the draws obtained for a stochastic variable, then these agents know that with sufficient repetitions, these draws will reflect the underlying parameters of that distribution. So this risky phenomenon can be treated as 'certainty-equivalent'. This is precisely the rationale that underlies the efficient-market hypothesis. There is risk at any point in time, but certainty over time; and never uncertainty.

12 These authors leave off the modifier 'mainstream'. By convention, mainstream economists making summative categorical statements of this type ignore the views of economists outside the mainstream.

13 See Dymski (2010) for references to these extensive literatures.

14 There is now even a small literature on equilibrium subprime lending; see Makarov and Plantin (2013).

15 See Milonakis and Fine (2009), especially chapters 12 and 14.

16 Ingham makes a parallel point: 'the relations between debtors and creditors are also the source of the systems fragility ... thus money cannot be neutral; it is the most powerful of the social technologies, but it is produced and controlled by specific monetary interests and is also inherently unstable' (2004, pp. 202–3).

17 See, for example, the Squam Lake report (French et al. 2010).

18 An important example is the folk theorem, wherein – in technical jargon – any feasible payoff profile that strictly dominates the minmax profile can be realized as a Nash equilibrium payoff profile, with sufficiently large discount factor. An application is to the famous 'prisoner's dilemma', wherein two players end up condemning one another to suboptimal outcomes due to their inability to signal a binding intention to cooperate. When this dilemma is repeated infinitely, then if the players are sufficiently patient, the cooperation outcome emerges as a sustainable outcome.

References

Aalbers, Manuel B. 2005. Who's afraid of red, yellow and green? Redlining in Rotterdam. *Geoforum* 36 (5): September, 562–80.

Aalbers, Manuel B. 2009. Geographies of the financial crisis. *Area* 41 (1):34–42.

Calomiris, Charles. 2014. *Fragile by Design: The Political Origins of Banking Crises and Scarce Credit*. Princeton: Princeton University Press.

Cassidy, John. 2008. The Minsky moment. *The New Yorker,* 4 February.

Christophers, Brett. 2011a. Follow the thing: money. *Environment and Planning D: Society and Space* 29:1068–84.

Christophers, Brett. 2011b. Credit, where credit's due. Response to 'Follow the thing: credit'. *Environment and Panning D: Society and Space* 29:1089–91.

Christophers, Brett. 2013. *Banking Across Boundaries: Placing Finance in Capitalism*. London: Basil Blackwell.

Christophers, Brett. 2015a. The limits to financialization. *Dialogues in Human Geography* 5 (2):183–200.

Christophers, Brett. 2015b. Financial crises.In *Wiley Blackwell Companion to Political Geography*, 1st edn, eds John Agnew, Virginie Mamadouh, Anna J. Secor and Joanne Sharp, 462–77. New York: John Wiley and Sons.

Christopherson, Susan, Gordon L. Clark and John Whiteman. 2015. Introduction: the euro crisis and the future of Europe. *Journal of Economic Geography* 15:843–53.

Clark, Gordon L., Maryann P. Feldman and Meric S. Gertler, eds. 2000. *The Oxford Handbook of Economic Geography*. Oxford: Oxford University Press.

Corbridge, Stuart. 1994. Plausible worlds: Friedman, Keynes, and the geography of inflation. In Corbridge, Martin and Thrift, pp. 63–90.

Corbridge, Stuart, Ron Martin and Nigel Thrift, eds. 1994. *Money, Power, and Space*. London: Basil Blackwell.

Corbridge, Stuart, and Nigel Thrift. 1994. Introduction and overview. In Corbridge, Martin and Thrift, pp. 1–25.

Davidson, Paul. 1972. *Money and the Real World*. London: Macmillan.

Davidson, Paul. 2015. What was the primary factor encouraging mainstream economists to marginalize post Keynesian theory? *Journal of Post Keynesian Economics* 37 (3):369–83

Diamond, Douglas. 1984. Financial intermediation and costly monitoring. *Review of Economic Studies* 51 (3):July, 393–414.

Dymski, Gary A. 1998. Disembodied risk or the social construction of credit-worthiness? In *New Keynesian Economics/Post Keynesian Alternatives*, ed. Roy Rotheim, 241–61. London: Routledge.

Dymski, Gary A. 2010. The international debt crisis. In *Handbook of Globalisation*, 2nd edn, ed. Jonathan Michie, 90–103. Cheltenham: Edward Elgar.

Dymski, Gary A. Forthcoming 2017. Finance and financial systems: Evolving geographies of crisis and instability. In *The New Oxford Handbook of Economic Geography*, eds Gordon Clark, Maryann Feldman, and Meric Gertler, ch. 28. Oxford: Oxford University Press.

Epstein, Gerald. 2006. Introduction: Financialization and the world economy. In *Financialization and the World Economy*, ed. Gerald Epstein, 3–29. Cheltenham: Edward Elgar.

Farmer, Roger. 1993. *The Macroeconomics of Self-Fulfilling Prophecies*. Cambridge: MIT Press.

Fox, Justin. 2009. *The Myth of the Rational Market*. New York: HarperBusiness.

French, Kenneth R., Martin N. Baily, John Y. Campbell, John H. Cochrane, Douglas W. Diamond, Darrell Duffie, Anil K. Kashyap, Frederic S. Mishkin, Raghuram G. Rajan, David S. Scharfstein, Robert J. Shiller, Hyun Song Shin, Matthew J. Slaughter, Jeremy C. Stein, and Rene M. Stulz. 2010. *The Squam Lake Report: Fixing the Financial System*. Princeton: Princeton University Press.

French, Shaun, Andrew Leyshon and Nigel Thrift. 2009. A very geographical crisis: the making and breaking of the 2007–2008 financial crisis. *Cambridge Journal of Regions, Economy and Society* 2:287–302.

Fujita, Masahisa, and Paul Krugman. 2004. The new economic geography: Past, present and the future. *Papers in Regional Science* 83 (1):139–64.

Gilbert, Emily. 2011. Follow the thing: credit. Response to 'Follow the thing: Money'. *Environment and Planning D: Society and Space* 29:1085–8.

Harvey, David. 1982. *The Limits to Capital*. Oxford: Basil Blackwell.

Harvey, David. 2003. *The New Imperialism*. Oxford: Oxford University Press.

Harvey, David. 2006. Neo-liberalism as creative destruction. *Geografiska Annaler Series B-Human Geography* 88B (2):145–58.

Harvey, David. 2014. *Seventeen Contradictions and the End of Capitalism*. London: Profile Books.

Ingham, Geoffrey. 2004. *The Nature of Money*. Cambridge: Polity Press.

Ingrao, Bruna, and Giorgio Israel. 1990. *The Invisible Hand: Economic Equilibrium in the History of Science*. Cambridge: MIT Press.

Jevons, Stanley. 1875. *Money and the Mechanism of Exchange*. New York: D. Appleton and Co.

Keynes, John Maynard. 1936. *The General Theory of Employment, Interest, and Money*. London: Macmillan.

Keynes, John Maynard. 1937. The general theory of employment, *Quarterly Journal of Economics* 51 (2): February, 209–23.

Krippner, Greta. 2005. The financialization of the American economy. *Socio-Economic Review* 3 (2):173–208.

Krippner, Greta. 2011. *Capitalizing on Crisis: The Political Origins of the Rise of Finance*. Cambridge, MA: Harvard University Press.

Krugman, Paul. 2009. How did economists get it so wrong? *New York Times*, 9 September.

Langley, Paul. 2015. *Liquidity Lost: The Governance of the Global Financial Crisis*. Oxford: Oxford University Press.

Laulajainen, Risto I. 2000. The regulation of international finance. In Clark, Feldman and Gertler, pp. 215–29 .

Leyshon, Andrew. 2011. Toward a non-economic, economic geography? From black boxes to the cultural circuit of capital in economic geographies of firms and managers. In Andrew Leyshon, Roger Lee, Linda McDowell and Peter Sunley, eds, 383–97.

Leyshon, Andrew, Roger Lee, Linda McDowell and Peter Sunley, eds. 2011. *The SAGE Handbook of Economic Geography*. Los Angeles: Sage Publications.

Makarov, Igor, and Guillaume Plantin. 2013. Equilibrium subprime lending. *Journal of Finance* 68 (3):June, 849–79.

Malinowski, Bronislaw. 1926. *Crime and Custom in Savage Society*. New York: Harcourt, Brace & Co.

Mann, Geoff. 2010a. Hobbes' redoubt? Toward a geography of monetary policy. *Progress in Human Geography* 34 (5):601–25.

Mann, Geoff. 2010b. Value after Lehman. *Historical Materialism* 18:172–88.

Mann, Geoff. 2013. The monetary exception: Labour, distribution and money in capitalism. *Capital and Class* 37 (2):197–216.

Mann, Geoff. 2015. Poverty in the midst of plenty: Unemployment, liquidity, and Keynes's scarcity theory of capital. *Critical Historical Studies* 2 (1), Spring:45–83.

Marshall, Alfred. 1920. *Principles of Economics*. London: Macmillan.

Martin, Ron. 1994. Stateless monies, global financial integration and national economic autonomy: The end of geography? In Corbridge, Martin and Thrift: pp. 253–78,

Martin, Ron. 2011. The local geographies of the financial crisis: from the housing bubble to economic recession and beyond. *Journal of Economic Geography* 11:587–618.

Martin, Ron. 2015. Rebalancing the spatial economy: The challenge for regional theory. *Territory, Politics, Governance* 3 (3):235–72.

Marx, Karl. 1906. *Capital, Volume 1: A Critique of Political Economy*. New York: Modern Library.

Mian, Atif, and Amir Sufi. 2014. *House of Debt*. Princeton: Princeton University Press.

Milonakis, Dimitris, and Ben Fine. 2009. *From Political Economy to Economics*. London: Routledge.

Minsky, Hyman P. 1978. The financial instability hypothesis: A restatement. *Thames Papers on Political Economy* 78 (3):1–24.

Minsky, Hyman P. 1986. *Stabilizing the Unstable Economy*. New Haven: Yale University Press.

Minsky, Hyman P. 1996. Uncertainty and the institutional structure of capitalist economies. *Journal of Economic Issues* 30 (2), June: 357-69.

Mirowski, Philip. 1984. Physics and the 'Marginal Revolution'. *Cambridge Journal of Economics* 8:361–79.

Mirowski, Philip. 1991. *More Heat than Light*. Cambridge: Cambridge University Press.

Mirowski, Philip,. 2013. *Never Let a Serious Crisis Go to Waste: How Neoliberalism Survived the Financial Meltdown*. London: Verso.

Parsons, Talcott. 1966. *Societies: Evolutionary and Comparative Perspectives*. Englewood Cliffs, NJ: Prentice-Hall, Inc.

Peck, Jamie. 2000. Doing regulation. In Clark, Feldman and Gertler, pp. 61–80.

Pryke, Michael. 2011. Geographies of economic growth II: Money and finance. In Andrew Leyshon, Roger Lee, Linda McDowell and Peter Sunley, pp. 286–302.

Radcliffe-Brown, Alfred R. 1958. *Method in Social Anthropology*. Chicago, University of Chicago Press.

Samuelson, Paul A. 1947. *Foundations of Economic Analysis*. Cambridge: Harvard University Press.

Samuelson, Paul A. 1948. *Economics*. New York: McGraw-Hill.

Sheppard, Eric. 2000. Teography or economics? Conceptions of space, time, interdependence, and agency. In Clark, Feldman and Gertler, pp. 99–119.

Shiller, Robert. 2008. *The Subprime Solution*. Princeton: Princeton University Press.

Skaggs, Neil. 1999. Changing views: Twentieth-century opinion on the banking school–currency school controversy. *History of Political Economy* 31 (2):361–91.

Stockhammer, Engelbert 2005. Shareholder value orientation and the investment-profit puzzle. *Journal of Post Keynesian Economics* 28 (2), Winter:193–215.

Streeck, Wolfgang. 2014. *Buying Time: The Delayed Crisis of Democratic Capitalism*. London: Verso.

Tarshis, Lorie. 1947. *The Elements of Economics*. Berkeley: University of California Press.

Tickell, Adam. 2000. Finance and localities. In Clark, Feldman and Gertler, pp. 230–47.

Tobin, James. 1972. Inflation and unemployment. *American Economic Review* 62 (1):1–18.

Williamson, Oliver. 1994. Transaction cost economics and organization theory. In *Handbook of Economic Sociology*, Neil J. Smelser and Richard Swedberg, 77–107. Princeton: Princeton University Press.

Wójcik, Dariusz. 2013. The dark side of NY–LON: Financial centres and the global financial crisis. *Urban Studies* 50 (13), October:2736–52.

Wolf, Martin. 2008. The end of lightly regulated finance has come far closer. *Financial Times*, 16 September 16.

Wyly, Elvin K., Mona Atia, Holly Foxcroft, Daniel J. Hammel and Kelly Phillips-Watts. 2006. American home: Predatory mortgage capital and neighbourhood spaces of race and class exploitation in the United States, *Geografiska Annaler* 88 B:105–32.

Part II
Financial Practices

5

Banks in the Frontline

Assembling Space/Time in Financial Warfare

Marieke de Goede

Money is always and everywhere the sinews of war: terrorists need vital funds to buy weapons, vehicles and arsenals. International action to counter terrorist financing is a bastion of peace and security in the world. (Michel Sapin, French Minister of Finance, in the wake of the Charlie Hebdo attacks, January 2015)[1]

Introduction: Banks on the Frontline

In February 2015, Paris-based Financial Action Task Force (FATF) released a report on the funding of Islamic State in Iraq and the Levant (ISIL). Created in the 1990s as an inter-state forum to tackle money laundering, the FATF was given a new life when it was charged with pursuing terrorism financing after 9/11. Formally, compliance with FATF recommendations is voluntary and best-practice-based. However, in practice, FATF members work hard to achieve positive scores and reports on their activities from this increasingly important forum (Heng and McDonagh 2008; Huelsse and Kerwer 2007). FATF has incorporated extensive terrorism financing recommendations among its money laundering guidance to states, and the aim of the report about ISIL was to design further strategies for 'disrupting and dismantling' ISIL financing (FATF 2015, p. 32). The report advocates active use of financial sanctions against (suspected) foreign fighters and disrupting financial flows to ISIL territories. Banks and financial institutions are required to monitor and report funds potentially linked to ISIL and to the travel of foreign fighters.

Money and Finance After the Crisis: Critical Thinking for Uncertain Times,
First Edition. Edited by Brett Christophers, Andrew Leyshon and Geoff Mann.
© 2017 John Wiley & Sons Ltd. Published 2017 by John Wiley & Sons Ltd.

Geographical markers play an important role in the FATF document: while the report does not formally propose monitoring or reducing transactions to Syria and Iraq as territorial entities, banks are warned especially of transactions to and from 'conflict zones or areas where ISIL operates' (FAFT 2015, p. 19). All seventeen anonymized 'case study' examples in the report discuss examples of money transfers, travel and cash-carrying to Syria, Saudi-Arabia and Turkish border regions. Case examples in the report place special emphasis on the dangers of funds being diverted from charitable collections (in, for example, Italy and Canada) to 'aid terrorists and their families' in the Syria–Iraq–Turkey border regions (FATF 2015, p. 20). The report concludes by noting the 'diverse geographic scope of the terrorist financing risk' associated with ISIL, and the need to develop further strategies to 'stop, restrain and enable confiscation of cash when reasonable grounds for suspicion exist of terrorist financing' (FATF 2015, p. 40).

Two elements of the FATF report frame the more general concerns of this chapter. First, the report positions banks and financial institutions on the frontline in the fight against ISIL. Banks have to detect and disrupt transactions possibly related to ISIL oil and antiques smuggling, as well to financial support from family and friends to individuals in contested territories in Syria and Iraq. This gives banks new complex responsibilities to freeze and report funds. Second, the spatial imaginaries in the report grapple with what is thought of as a dispersed and diffuse threat. While notions of danger in the war on terrorism are certainly territorial (Elden 2007) – focusing explicitly on Iraq, Syria, Turkey – they are also constitutive of new, networked, understandings of danger (Coward 2009; 2013). The FATF report presents a complex geography in which charitable donations in Canada are thought to support violence in Syria; and ATM withdrawals at the Turkish-Syria border indicate potential travel of (what are now commonly called) foreign fighters. The FATF report on ISIL funding seeks to outline a strategy to filter and interrupt potentially suspicious transactions to a diffusely defined border region and along specific travel routes. What's left implicit in the report, however, is its equally important objective of continued facilitation of financial flows to other parts – and with other parties – of Syria and adjacent territories. In this sense, the politics of the FATF report is one of defining 'good' and 'bad' circulation, which is how Foucault typified the practice of *security*. Instead of a strategy focused on 'fixing and demarcating the territories', the strategy of the FATF is one 'of allowing circulations to take place … controlling them, sifting the good and the bad, ensuring that things are always in movement … but in such a way, that the inherent dangers of this circulation are cancelled out' (Foucault 2007, p. 65; also Amoore and de Goede 2008).

The 2015 FATF report follows a longer but relatively invisible intensification of what has been called 'financial warfare' (Zarate 2013) or the 'weaponisation of finance' (Holodny 2015). Juan Zarate, former US assistant secretary of the Treasury, calls countering terrorist and illicit financial flows a 'new form of warfare', whereby private-sector interests and the banking sector are leveraged to 'uncover the financial footprints and relationships of the terrorist network' (2013, p. 29). For Zarate, it is a way to harness the power of US financial institutions in the service of security interests and the global war on terror. Financial warfare positions banks and financial institutions on the frontline of security practice and fighting terrorism. It carves out a new role for money and finance in security, whereby money comes to be considered as a 'tool of combat' (Gilbert 2015a, p. 205; also Gilbert 2015b). In this sense, financial warfare profoundly impacts financial practices, and has the capacity to shift banks' risk calculations and reroute international money flows.

This chapter examines the interconnections between finance and security in the post-9/11 context, and analyses the novel spatial configurations that have accompanied the pursuit of terrorist monies. I deploy the term finance/security *assemblage* to analyse the uneasy alliances between banks and financial institutions on the one hand, and security authorities including intelligence agencies and police on the other (building on de Goede 2012). Countering terrorism financing does not so much erect new national borders within the global juggernaut of financialization – as it has sometimes been understood (Biersteker 2004) – but enacts reconfigurations of regulatory space/time, that impact financial practices and enable security decisions in particular ways.

This chapter focuses on the spatiality of financial warfare, and analyses the way in which US authorities have rearticulated their jurisdictional reach in recent investigations to encompass *all* dollar-denominated transactions globally. Luiza Bialasiewicz and colleagues have suggested that security spaces can be understood as 'imaginative geographies', articulated, carved out and performed 'by a variety of security advisers and popular academic commenters' (Bialasiewicz et al. 2007, p. 409). The production of space, in this analysis, functions through 'citational practices' whereby state-based geographical imaginaries resonate with popular culture and academic commentary (Bialasiewicz et al. 2007, p. 409). Space, in this sense, involves the work of *spacing*: it is 'an effect of practices of representation, valorization and articulation' (Gregory 2004, p. 19; also Gregory and Pred 2007). This chapter hones in on the recent investigation of British banking conglomerate HSBC by the US Senate and the US Department of Justice, and interrogates the performativity of space in this case. It argues that we can observe the emergence of a transactional, rather than territorial, understanding of US regulatory space. The new geographies of financial warfare do not just have an impact on banks' compliance practices, but also influence banks' risk calculations concerning un/viable transactions, clients and business lines. The chapter starts with an analysis of post-9/11 financial warfare as the development of new spatio-temporal modes of security intervention, and puts forward the notion of 'assemblage' as a helpful analytical starting point. It then goes on to analyse in some depth the US Senate investigation of HSBC, before focusing on recent debates concerning preemptive account closures in the UK.

Finance/Security Assemblage

Though the political urgency of pursuing suspect money flows and potential financing of terrorism has increased significantly since September 2001, its juridical history predates the 9/11 attacks. In 1999, the United Nations adopted the Convention on the Suppression of the Financing of Terrorism, which obliges member states to criminalize the act of 'terrorist financing', defined in a broad sense as the wilful provision or collection of funds or assets ('of every kind')

> with the intention that they should be used or in the knowledge that they are to be used, in full or in part, in order to carry out ... Any ... act intended to cause death or serious bodily injury to a civilian ... when the purpose of such act, by its nature or context, is to intimidate a population, or to compel a government or an international organization to do or to abstain from doing any act.[2]

In this manner, the Convention developed a de facto definition of terrorism –which had been controversial at the UN throughout the 1970s and 1980s. The UN Convention is said to have created a veritable 'paradigm shift' in its redefinition of terrorism *beyond* violent acts, enabling a broad criminalization of financing and facilitating acts (Lehto 2009). Financial institutions and intermediaries were newly charged with identifying transactions potentially linked to terrorism, policing remittances and judging the legitimacy of wire transfers or ATM transactions. If the pursuit of money laundering focused on big cash amounts and post-crime transactions, the new responsibilities for banks and other financial institutions in the context of pursuing terrorism financing focus on identifying small, mundane transactions *potentially* linked to future crimes (Malkin and Elizur 2002). In other words, the focus of banks and regulators is increasingly on relatively small transactions, understood to be 'pre-crime' (Zedner 2007).

When, in the immediate aftermath of 9/11, transnational money laundering and counter-terrorism financing regulation were propelled to the top of policy agendas, some observers celebrated a renewed state sovereignty in the globalizing marketplace. Thomas Biersteker (2004, p. 73), for example, welcomed the 'sea change in the tolerance of financial re-regulation across the globe', while William Vlcek (2008) declared a 'Leviathan Rejuvenated'. However, a focus on the reassertion of sovereign statehood underplays the spatio-temporal complexities of the war on terrorism financing. Instead of an unequivocal resurgence of state sovereignty in the regulatory landscape, we observe new 'assemblages', through which administrative governmental power is allied with private companies. This includes, for example, placing banks and other private money transfer companies like Western Union and SWIFT on the frontline of security practice. These companies are both newly empowered to deem transactions suspicious and to block them, and uneasily positioned as quasi-official security agencies.

Elsewhere, I argue that countering terrorism financing is perhaps less about cutting off financial flows to terrorists, and more about enabling new spatio-temporal strategies of security intervention (de Goede 2012). The effectiveness of countering terrorism financing is disputed and difficult to measure (e.g., Deloitte 2011, p. 95; Levi 2010). Records on disrupted plots and monies seized are usually not public. Very recently, FATF has started to use successful court convictions as a measure of effectiveness when evaluating member states. However, despite its contested efficacy, very real power mechanisms are at work in the name of countering terrorism financing and financial warfare more broadly. For the UK Treasury (2007, p. 10), for example, 'Financial information has a key intelligence role', because it allows law enforcement to:

- *look sideways*, by identifying or confirming association between individuals and activities linked to conspiracies, even if overseas – often opening up new avenues of inquiry; and
- *look forward*, by identifying the warning signs of criminal or terrorist activity in preparation (emphasis in original).

The effect of this logic is to extend the security apparatus into the domain of everyday life by broadening the *space* of security (looking sideways) and extending the *time* of security (looking forward). The war on terrorism financing broadens the space of security because it can criminalize associations, links or relations. It moves the 'lines of sight' of security practices toward the support networks of potential terrorists (Amoore 2009), including a persistent and controversial focus on charitable organizations and donations thought to hold propensity of supporting violence and radicalization (Howell and Lind 2009). In addition, pursuing terrorist monies extends security across *time* because it enables preemptive action on the basis irregular risk profiles and suspect networks. For example, FATF (2015, p. 40) calls on its member states to create 'legal and operational frameworks to stop, restrain and enable confiscation' of financial flows 'when reasonable grounds for suspicion exist of terrorist financing'. The notion of reasonable suspicion here functions to enable the identification and disruption of transactions that are inscribed with the intention to support potential future violence.

If the pursuit of terrorist monies creates new spaces and new time horizons for security intervention, what does this mean for our understanding of financial geography and power? Emily Gilbert (2013, p. 56) has raised the question of the spatial dimension of the pursuit of terrorist monies: 'Are the geographies that emerge simply reinforcing predictable characterizations of risky geographies?... Are all countries equally interested and capable of exerting extra-territoriality? How are other countries caught up in the regulatory net of [the] US [or the] EU?' Gilbert raises poignant questions about how financial warfare influences, shapes and reorients financial geographies of profit, risk and regulation. Conventionally, the geography of power in this domain is thought to emanate from the US. American initiatives concerning asset freezes, confiscations and regulations have been incredibly important in global counter-terrorism financing agendas (e.g., Biersteker and Eckert 2007; Warde 2007). However, this chapter proposes a shift in perspective, for at least two reasons. First, though we can safely say that the power of financial warfare overwhelmingly emanates from the US (but also the EU), this still says little about the configurations of power *inside* the US. As a focus on the HSBC case below shows, jurisdictional reach is not simply exercised, but is reworked and reconfigured in the name of countering terrorism financing. In a sense, we may say that the 'US' is itself a mobile configuration – elements of which become (dis)empowered (Allen and Cochrane 2007; 2010). In this context, the power of the US Senate and the relatively little-known US Treasury Office of Foreign Assets Control (OFAC) are critically interrogated. Second, the power of financial warfare is enacted through complex alliances with the private sector, as has been noted above. This has important implications for how we understand geographies of power: freezing, reporting and confiscating are ultimately done by mid-level financial bureaucrats and within the private, commercial spaces of banks and financial institutions. These private actors do not simply *implement* policies, but enact them in ways that foster new directions and graft on novel (commercial) goals (Amicelle and Jakobsen 2016).

To start addressing the complex geographies of power in financial warfare, we can draw upon a rich literature that theorizes financial practices as specific forms of time/space 'imaginaries' (Pryke and Allen 2000). For Pryke and Allen, financial instruments such as derivatives should be understood as 'monetized time/space' because they commercialize temporal uncertainties and geographical differences, allowing companies to hedge them (or speculators to bet on them) (2000, p. 282). Put simply, derivatives render the uncertain future tradeable in the present: their rationale is to 'help make the future both profitable and secure' (Pryke and Allen 2000, p. 276). Financial instruments render liquid the potential price fluctuations of certain referenced securities or currencies across time and space. Derivatives make present possible futures: their objective is not to *predict* the future, but to generate a multiplicity of possible futures that are rendered liquid (and thus tradeable) in the present (Amoore 2011; Arnoldi 2004; Cooper 2010). As one of the 'fathers' of option pricing theory explains the reasoning behind his models for derivative valuation: '[T]he future is uncertain...and in an uncertain environment, having the flexibility to decide what to do after some of that uncertainty is resolved definitely has value. Option-pricing theory provides the means for assessing that value' (Merton 1998, p. 339). Thus, derivatives grant financial investors flexibility in the face of uncertainties (Tellmann 2014). For Pryke and Allen (2000, p. 282) this involves a 'recoding of time–space' that amounts to a 'new monetary imaginary'.

These literatures problematize notions of 'global finance' as a coherent and hierarchically superior locus of power. There are complex ties between everyday, local monies and globally elongated networks of financialization. These are not reducible to a separation of global financial macrostructures on the one hand, and intimate local monies on the other (also Leyshon and Thrift 2007). As Tickell points out, there is 'something of a paradox' at work when, on one hand, the literature draws attention to the ways in which derivatives 'flatten differences' and foster processes of financial market convergence, while, on the other hand, the same literatures draw attention to the enduring fact that 'people inscribe money with values and meanings that vary over time and space' (2003, pp. 118–19; also Leyshon 1997). For example, the intimate monies of remittance transfers tying diaspora families to their countries of origin, support a profitable and globe-encircling wire transfer industry, including companies like Western Union. According to World Bank estimates, recorded remittances to the developing world reached over US $400 billion in 2010, constituting 'a vital source of financial support that directly increases the income of migrants' families'.[3] As Viviana Zelizer (2005) has convincingly argued, understanding the 'complex economic web' of remittance flows requires attention to the physical and social earmarking of monies, as well as moral boundaries drawn around budgets in everyday lives. In other words, the familial obligations of remittance-sending migrants cannot just be reduced to the 'micro' element of a macroeconomic aggregate, but are an important part of this economic configuration. Similarly, Langley (2008, p. 24) emphasizes that researchers should resist 'the temptation to simply explain developments within "small" everyday networks in terms of an apparently "bigger picture" of "the financial system"'. Instead, what matters is whether and how financial practices are connected to longer or shorter financial networks 'that continually have to be worked upon' (Leyshon 1997, p. 389).

In addition, a large body of literature problematizes the assumption that offshore financial centres are geographically distant from the financial centres of London, New York or Tokyo. Though territorially distant and inscribed with cultural difference, the practice of 'offshore' is intimately connected to centres of global finance and the ways in which financial instruments are structured, negotiated and traded. As Susan Roberts (1994, p. 111) puts it succinctly, 'Offshore financial centres are at once at the margins and at the centre of global capitalism's displacement of crisis' (see also Aitken 2006; Tellmann and Opitz 2011). In short, there are elongations, approximations and accelerations at work in capitalist financial geographies – and these are insufficiently captured through contrasting micro and macro 'scales' (Allen and Cochrane 2007, p. 1167). Offshore finance illustrates the problems posed by thinking about finance by pitting centre *versus* margin.

I propose the notion of 'assemblage' as a way of conceptualizing and critiquing the uneven geographies of combating terrorism financing, and unpacking its power effects across geographical distances and policy scales (e.g., Bennett, 2005; Anderson and MacFarlane 2011; Acuto and Curtin 2013). According to Giorgio Agamben (2009, p. 3), an assemblage is defined through its heterogeneity, its strategic functionality and its operation at the intersection of power and knowledge (1990, p. 3). An assemblage, then, is understood as 'heterogeneous ... political formation' that is mobile, emergent and dispersed – but that nevertheless exercises considerable power in the name of its strategic functionality (Allen 2011, p. 154). Conceptually, it helps explain how the interplay of a heterogeneity of elements, including, for example 'regulatory decisions, laws, administrative measures ... [and] moral and philanthropic propositions', enables a certain strategic functionality and outcome (Foucault, quoted in Agamben 2009, p. 2). This interplay may at times lead to relatively stable formations and 'well-ordered coherent wholes' (Bueger 2013, p. 62). However, such stability and order can never be assumed or taken for granted, but needs to be itself explained. Thus, in my analysis of the strategic outcomes of the pursuit of terrorism financing, I am not necessarily arguing that policies have hidden agendas that deviate from their stated objective. On the contrary, it is the cumulative effect of the finance-security assemblage doing 'exactly what it says on the cover' that, taken together, produce these effects (Christophers 2012). This is what Jane Bennett (2005, p. 447) calls *the agency of assemblages*: 'the distinctive efficacy of a working whole made up, variously, of somatic, technological, cultural, and atmospheric elements. Because each member-actant maintains an energetic pulse slightly "off" from that exuded by the assemblage, such assemblages are never fixed blocks but open-ended wholes.' Put differently, outcomes are conceived as 'unstable cascade' rather than as causal effects (Bennett 2005, p. 457; also De Goede 2012: chapter 2).

The transnational landscape of laws, institutions, treaties and private initiatives that play a role in fighting terrorism financing, is complex, not necessarily transparent, and pulling in different directions. Considerable tensions, gaps and disjunctures persist within it, problematizing (too) coherent and powerful renditions of US hegemonic power or homogenous neoliberalism (e.g., Bennett 2010). In the war on terrorism financing, these tensions are important and long-standing. For example, within this regulatory complex a struggle persists over whether efforts should concentrate on *freezing* monies (with all its attendant juridical problems) or on *following* monies (deploying sophisticated financial data analysis to rendering visible suspect networks). A related tension – explored below – concerns the contingency of commercial opportunities made possible with fighting terrorism financing. Cleary, a strong transnational policy agenda to reregulate transnational banks and impose substantial compliance duties and fines is broadly understood as restricting the power of capital (rather than fostering it). At the same time however, important opportunities are emerging for (US-based) firms and financial data mining software companies.

The HSBC Settlement

In 2012, British-headquartered bank HSBC (Hong Kong and Shanghai Banking Corporation) concluded a settlement with US authorities, including the Department of Justice and the Office of Foreign Assets Control (OFAC), for 1.9 billion US$. The agreement followed a protracted investigation by the US Senate Committee on Homeland Security and Governmental Affairs into alleged money laundering and terrorist finance abuses at HSBC, and settled claims that the bank deliberately evaded US controls and sanctions. Under the terms of the settlement, HSBC acknowledges violations of the Bank Secrecy Act, the International Emergency Economic Powers Act and the Trading With the Enemy Act. The Senate Committee released a 300-page report, discussed at a public hearing in July 2012. It was announced as a landmark case for the US Department of Justice. As Assistant Attorney General Breuer said at the press conference: 'Today represents the unit's largest case to date … We intend to continue ensuring that financial institutions do their part to protect the US financial system against the threat of money laundering.'[4]

I use the HSBC settlement as a starting point, an 'individual [point] of leverage', to render visible a part of the complex and modulating finance/security assemblage that takes shape in the name of counter-terrorism financing (French et al. 2011, p. 812). The settlement is part of a general trend in which OFAC has asserted itself as an aggressive overseer of US sanctions policies (in particular in relation to the Iran sanctions), leading to truly massive fines for (mostly) European-based banks: ING paid US$619 million in 2012; Barclays, US$298 million in 2010; ABN Amro, US$500 million in 2010 and Credit Suisse, US$ 536 million in 2009. In 2014, BNP Paribas settled for an unprecedented US$8.9 billion over alleged evasion of sanctions against Iran.[5] I focus on the HSBC settlement as one lens on the geographies of contemporary financial regulation. By investigating the associations, analysing the debates, and uncovering the rationale of the settlement, we can bring into view the larger finance/security assemblage – not as a pre-existing structure, but as a modulating regulatory complex that itself requires explanation. When understood as part of a wider assemblage that fuses finance and security in the name of countering terrorism, we can see how the OFAC enforcement actions restrict and redirect global financial flows and structure financial risk practices.

I argue that the HSBC settlement and hearing develops and enacts an important and relatively novel spatial imaginary, that understands the reach of OFAC to be defined through what comes to be called 'US currency space'. The senators leading the investigation stressed the importance of US law in the context of anti-money laundering and terrorist financing, and the obligation of globally-operating financial institutions to obey US standards and sanctions. But, interestingly, the senators' notion of US law and proscription appeared *not* bound by US territory or conventional conceptions of territorial jurisdiction. Instead, they emphasized the application of US standards and sanctions to *all transactions denominated in US currency*. As Senator Levin (D-Michigan) pointed out, many financial transactions – in trade, investment and speculation – are dollar-denominated, through habit and calculative standardization. A trade transaction between Singapore and Germany, or between

Turkey and Japan, might be denominated in US currency. As he said in his opening statement: 'Global banks want access to US dollars because they are accepted internationally, they are the leading trade currency, and they hold their value better than any other currency' (Levin, US Senate 2012a, p. 1).

Jurisdictional space, in the formulation of the senators leading the HSBC hearing, is not territorial but transactional: it is defined through the flows and circulations of US-dollar denominated transactions. Here, US national jurisdictional boundaries are elongated and rendered flexible. The new geography of counter-terrorism financing regulations becomes the space of dollar-denominated transactions. It operates a form of power that (Foucault 2007, p. 65) calls *security*; and 'that is no longer that of fixing and demarcating the territory, but of allowing circulations to take place, of controlling them, sifting the good and the bad, ensuring that things are always in movement'.

In the Subcommittee's approach, a bank's use of US dollar-denominated transactions becomes a point of leverage to enforce US terrorist financing rules, even for banks that are not primarily US-based. In his opening statement Levin warned that a US-based bank office 'can become a sinkhole of [terrorist finance] risk for an entire network of bank affiliates and their clients around the world playing fast and loose with US rules' (US Senate 2012a, p. 2). In this manner, he suggests that such banking affiliates, however small in relation to the wider banking conglomerate, should function as point of leverage for OFAC and DoJ to enforce US standards (for example with regard to sanctions). As became apparent during the Senate hearing, HSBC functions not as a single corporate unit, but as a global confederation of banks, a disjointed network of local affiliates created by its policy of buying up national banking networks. According to David Bagley (2012, p. 2), head of compliance at HSBC, the bank's corporate structure is like 'an international federation of affiliates around the globe, many of [which] began as relatively small independent banks that HSBC acquired over the years'. Some of these national affiliates and local branches functioned quite independently, with so-called 'know-your-customer' procedures based less on global compliance practices, and more on personal contact and local knowledge. Part of the problem with Mexican branches of HSBC, for example, which allegedly served as conduits for drug money, appeared to be that local branches functioned independently of the Mexican head office, and had incentives to bring in business. As noted during the hearings, the Mexican local branches operated with 'a performance management system ... inherited from the former Bital bank that was heavily focused on business growth rather than control ... In addition ... Mexico was a data-poor environment, making it difficult to verify the identities of customers' (US Senate 2012a, p. 23). The local Mexican branches functioned with models of trust and personal contact in assessing money laundering risks that in most countries have been superseded by electronic systems for detecting suspicious or abnormal transactions (Leyshon and Thrift, 1999; Maurer 2005).

The point here is not to dispute the money laundering concerns regarding Mexican transactions, but to render visible the workings of an assemblage in which local Mexican bank branches are disciplined into operating with particular modes of assessing money laundering risk and into complying with US sanctions lists. One of the concrete outcomes of the hearings was HSBC's pledge to 'adopt and enforce the adherence to a single standard globally that is determined by the highest standard that we must apply anywhere' (US Senate 2012a, p. 51). It was also acknowledged, that often, 'this will mean adhering globally to US regulatory standards' – meaning that, effectively, US standards and sanctions become enforced in local Mexican banks, through private sector decisions at the HSBC London head office (Larner and Laurie 2010).

The extra-territorial enforcement of US standards is of particular interest in relation to the OFAC sanctions lists that interdict financial transactions with listed persons and entities, and complement other terrorism blacklists, including the UN Security Council al-Qaeda targeted sanctions lists and the EU terrorism blacklist (de Goede 2011; de Goede and Sullivan 2016). OFAC, established in 1950 to block Chinese and North Korean assets held within US jurisdiction, has come to play an important yet largely invisible role in the war on terror. Pursuant to Executive Order 13224 of September 2001, OFAC delivers targeted sanctions against named individuals and entities that are not always related to national sanctions programmes against so-called 'rogue states' like Iran and North Korea. The OFAC Specially Designated Nationals List plays an important role in the US wars on drugs and terror, and includes 'individuals, groups, and entities, such as terrorists and narcotics traffickers designated under programs that are not country-specific'.[6] What was novel about sanctions lists after 9/11, including OFAC, but also UN and EU lists, is their focus on individuals and groups not tied to particular territories, but connected to al-Qaeda and diffuse terrorist networks (Sullivan 2015). Nevertheless, significant differences remain in the precise composition of sanctions lists in different regions and, in particular, on the two sides of the Atlantic. There has been significant resistance from the European Court of Justice in recent years on the legitimacy of sanctions measures, for example in the case of Iran. In 2013, the ECJ annulled the sanctions against a number of Iranian companies and banks, partly because the European Council (which formally takes the sanctions decision in the EU) 'has not proved the facts of which it accuses those four companies'.[7]

The HSBC investigation by the US Senate includes detailed documentation of alleged circumvention of OFAC prohibitions by HSBC affiliates in Europe and the Middle East in transactions with Iran, Burma, Cuba, North Korea and the Sudan. For example, and as acknowledged in the settlement, HSBC cleared US dollar-denominated transactions for Iranian Bank Melli through a complex construction that made the transactions appear legitimate. It appeared that European-based affiliates of HSBC discussed with Bank Melli exactly how to fill in the required money transfer forms so that the transactions would not be subject to an OFAC filter check in the US. The complex details of the allegations are not relevant to our discussion. Rather, what is important are the differences in interpretation between the US senators and the HSBC representatives. While the Senate Committee interpreted this process as a deliberate attempt by the European affiliates to circumvent US standards and misinform the US-based HSBC affiliate, the HSBC head office claimed it was 'ensuring that the payments were... compliant [with US regulation] before they got to the United States' (US Senate 2012a, p. 36). The point of drawing out these interpretive differences is not to take sides between OFAC and HSBC, but to render visible the complex interpretive practices at work in securing financial circulations.

The geography of the finance/security assemblage thus is clearly one in which US power is exerted extraterritorially, as analyses of US power over capital markets would lead us to expect (e.g., Helleiner 1994; Germain 1997). However, rather than endorsing a generalized notion of US hegemony, I have drawn attention to the specific nature of this assemblage, which involves the little-known administrative office of OFAC (housed in the US Treasury), and operates not through state-to-state pressure but through fines and investigations of private commercial institutions operating within a broadly defined 'US-currency area'. The specific nature of this assemblage – working through administrative rules and the private sector – is important to its functioning: it involves the operation of power which is not strictly territorial but which is transactional. This power, importantly, works through the decisions of private sector institutions, where security objectives become grafted onto business concerns. Indeed, it works intentionally to *circumvent* negotiation and bargaining at the nation-state level, especially regarding sensitive political decisions to blacklist institutions and sanction individuals.

Consider the example of transactions with Bank al-Rahji, which played an important role in the allegations that HSBC circumvented OFAC prohibitions (US Senate 2012b, p. Chapter IV). Al-Rahji is one of the largest private Saudi Arabian banks, and according to the Senate, a key founder of the bank 'was an early financial benefactor of al-Qaeda and … provided accounts for suspect clients' (US Senate Hearing 2012a, p. 5). The Senate investigative report develops this claim with reference to the so-called 'Golden Chain' list of alleged financial backers of al-Qaeda. This list was found in 2002 in the offices of a Bosnian charity, the Benevolence International Foundation (BIF), itself accused of being a front for an al-Qaeda financing network (US Senate 2012b, pp. 194–5). The Report details a host of allegations against al-Rahji, based on information from the CIA and the *Wall Street Journal*, including its alleged support of extremists in Indonesia and Afghanistan (US Senate 2012b, pp. 196–8; Simpson 2007). Al-Rahji became a client of HSBC when the conglomerate bought Republic Bank of New York, where al-Rahji banked since the 1970s (US Senate 2012b, p. 203). Though HSBC classified al-Rahji as a high risk client, the relationship continued.

What is important to establish for my argument concerning the way in which the finance-security assemblage operates is not whether the accusations against al-Rahji are true or false, but first, that al-Rahji bank is not included in European, nor on the UK Treasury sanctions lists, where the HSBC transactions took place; and, second, that the accusations of terrorism financing by and through al-Rahji were not established before a court of law. Al-Rahji contested the allegations made in the *Wall Street Journal* and based on the Golden Chain list in a 2003 libel case in the UK. The case came before the British High Court, which found in favour of al-Rahji and wrote in its judgement that, 'The Golden Chain … is *said* to be a list of donors to Al-Qaeda … [But it does not] emerge clearly what the meaning of the document is – whether, for example, it purports to be a list of donors or a list of those who might be approached for funding. Nor is it clear who created the document or when it came into existence' (High Court 2003, §23, emphasis in original). The *Wall Street Journal* was ordered to amend its article and published a letter by the al-Rahji Chief Executive Director in 2004.[8] The Senate investigative report does acknowledge that 'neither the bank nor its owners have ever been charged in any country with financing terrorism' (US Senate 2012b, p. 202). Nevertheless, the Senate sought to enforce the prohibition of dealing with al-Rahji extraterritorially.

When reinserted into this longer historical narrative, we can understand the fine imposed on HSBC as a novel mode of leveraging US currency power in order to take action against the Saudi bank, effectively circumventing the UK libel Court decision. The novel imaginary of jurisdiction as transactional rather than territorial, pressured HSBC to sever business ties with a bank that was not on the EU sanctions list in 2010. The HSBC investigation enforced the prohibition against dealing with al-Rahji extraterritorially, even though the bank was not sanctioned in Europe or the Middle East, and helped cut it off from global financial circulations.

A final element to be drawn out about the finance/security assemblage, which, as we have seen, operates through extraterritorial administrative measures, is its market power. The pursuit of terrorist monies and active interference with banking activities can easily be seen as a constraint on market power and a reassertion of the state – especially because it signalled a 'U-turn' for the Bush government, which prior to 9/11 pledged to resist global money laundering regulation (Warde 2007). However, this is not simply a state-versus-market story. A substantial market in professional compliance services, companies and software is fostered through Senate and OFAC actions. The market in compliance software, that helps banks mine for suspicions transactions and fulfil regulatory obligations, has grown significantly in recent years, and is expected to exceed US$10 billion in the coming two years (Keatinge 2014, p. 50). One of the Senate recommendations was that the HSBC US affiliate increase Anti-Money Laundering (AML) resources, 'hire qualified staff … and implement an effective AML monitoring system for account and wire transfer activity … including OFAC alerts' (US Senate 2012a, p. 12). During the hearing, it became apparent that HSBC has since contracted financial technology firm Norkom to deliver compliance software to fulfil these obligations. As the new chief executive of HSBC Bank USA put it: 'We have installed NORKOM. It has been a huge investment' (US Senate Hearing 2012a: 65). Norkom is one of the market leaders in software compliance products and was bought by British defence multinational BAE systems for €271million in 2011 (Brown 2011).

Risk and Preemptive Account Closures

The HSBC investigation and settlement is a window onto the workings of a finance/security assemblage, in which jurisdictional reach became understood as transactional instead of territorial, through the notion of the US currency domain. As mentioned, the HSBC case was one of a number of settlements between OFAC, the US Justice Department and European banks. By following the leads of the HSBC case further, we can see how the recent OFAC fines 'resonate' across global banking assemblages, reorient risk assessments and affect in very specific ways banking access and international remittances. In short, the HSBC case eventually led to great societal concerns over banking access of migrant and vulnerable client groups, as this section discusses.

Banks and other financial institutions are required to undertake terrorism financing risk assessments on the basis of what is called a 'risk based approach'. This means that the regulator does not prescribe particular categories and thresholds of suspicion, but authorizes banks to undertake in-house risk assessments and subjective judgements. This approach works with the assumption that banks themselves know best what is unusual and abnormal within their systems, and it encourages financial institutions to remain 'as supple as the criminals and terrorists themselves' (UK Treasury 2007, p. 13; De Goede 2012: chapter 3). Counter-terrorism financing regulation purposefully allows banks and other financial institutions substantial *discretion* in implementation, to encourage them to remain vigilant and pro-active. According to Faravel-Garrigues and colleagues (2008, p. 2), 'banking institutions enjoy a wide room for maneuver' and have become 'sentinels' for police and security work.

Often, banks' in-house risk assessments rely on datasets and analytical software tools developed by commercial vendors, for example Norkom, Thomson Reuters and Fiserv. The ways in which knowledge is generated concerning suspicious transactions inside banks, combine software-generated 'red flags' with situated judgements based on 'personal regard and ethical scrutiny' (Maurer 2005, p. 477). In this process, potential terrorist financing risk indicators commingle with other types of high-risk transactions and clients, including money laundering risk indicators, and the risks associated with banking so-called Politically-Exposed Persons (PEPs). This also suggests that judgement concerning suspicions transactions and terrorism financing risk is inseparable from *practice* (Boltanski and Thévenot 2000). For example, as Anthony Amicelle and Elida Jakobsen show in their ethnographic study of banking practices, banks appropriate sanctions lists in particular ways. Sanctions lists, in their analysis, 'acquire multiple, simultaneous identities in the course of [their] banking appropriation (from a compliance and disciplinary device to an inclusionary and exclusionary device)' (2016, p. 30). In short, business logics and commercial objectives become 'grafted onto' the instruments that are supposed to detect terrorist finance or money laundering risks (Li 2007).

During the Hearings, representatives of HSBC worked very hard to satisfy the Senate that they were aware of terrorist financing and money laundering risks in their organization and prepared to remedy past mistakes. One of the problems at HSBC, it appeared, was a backlog of 'over 17,000 alerts' generated in (automated) compliance systems that required further review in order to assess whether these should lead to suspicious transactions reports, freezing decisions or account closures (US Senate 2012b, p. 3). According to the subcommittee, HSBC suffered from 'poor procedures for assigning country and client risk ratings … inadequate and unqualified AML staffing; inadequate AML resources; and AML leadership problems' (US Senate 2012a, p. 3). In response,

HSBC argued that it is working hard implement new compliance procedures and software systems. Irene Dorner, the new Chief Executive of HSBC US testified that, since the investigation, the bank had 'exited, as a result of rolling out this remediation, in the order to 14,000 customers because they simply did not fit our risk appetite' (US Senate 2012a, p. 55). However, this rather offhand remark points to much larger – and much disputed – shifts in bank risk assessment practices. The objective of risk-based regulation is that banks and financial institutions develop fine-grained models of risk assessment – differentiating risk profiles for geographical areas, account types and client groups, for example. However, what we have seen in recent years – in conjunction with the financial crisis as well as the OFAC fines – is that some banks have decided to exit particular client relationships altogether, on the basis of what they call their 'risk appetite' (Durner and Shretret 2015; Keatinge 2014).

In July 2014, HSBC announced it would be closing the accounts of a number of UK-based Muslim groups, including the London Finsbury Park Mosque, the Cordoba Foundation and Ummah Welfare Trust. Muslim groups protested the decisions and raised concern over religious discrimination. Finsbury Park Mosque chairman Mohammed Kozbar told the BBC, 'For us it is astonishing – we are a charity operating in the UK, all our operations are here in the UK and we don't transfer any money out of the UK' (in Laurie 2014). Mr Al Tikriti, of the Cordoba Foundation, complained the bank did not inform him of the reasons for termination, noting he had been with the bank since the 1980s and 'has rarely been overdrawn' (in Laurie 2014). However, HSBC denied allegations of religious discrimination, stating that the decisions were based on a general review of its client base, which concluded that 'provision of banking service ... now falls outside our risk appetite', but offering no further explanation or a right to appeal the decision (as quoted by Oborne 2015; also Barrett 2014).

The HSBC account closures in 2014 followed an earlier British debate surrounding Barclays Bank's decision, in 2013, to discontinue partnerships with eighty Money Service Businesses (MSBs) remitting money to Somalia. There were no allegations of fraud or misuse of the accounts. On the contrary: the companies were considered 'model customers of Barclays'.[9] However, and partly motivated by the fines levied against HSBC and other European banks, Barclays conducted an internal risk review and decided to exit these relationships because of a 'perceived higher level of risk' in the small-scale MSBs sector.[10] The account closures by HSBC and Barclays can be understood as *preemptive* because they were not motivated by past misuse, but to avoid potential future abuse. As Claudia Aradau and Rens van Munster (2007) have shown, preemption addresses dangers where there is a large measure of uncertainty in risk calculation, but where the consequences of not acting could potentially be catastrophic (also de Goede and Randalls 2009). According to Barclays, the legitimacy of transfers from these accounts to Somalia and Eritrea could not be fully assessed. This became pressing in a context in which anti-terrorism financing requirements oblige banks to report transactions and freeze monies associated with the Somali terrorist network al-Shabaab. In a public statement, Barclays said that it 'does not want to unwittingly facilitate such transactions, given these serious risks'.[11] Moreover, the bank explicitly referred to the OFAC fines levied on HSBC and other banks, and wrote: 'if we were caught up in such transactions Barclays could be punished by our regulators and potentially fined, as we have seen with global banks receiving fines of hundreds of millions for anti-financial crime failures'.[12]

The Barclays decision was hotly debated in Britain, with prominent voices like Olympic gold medallist Mo Farah amongst others calling on the bank to reconsider (Muir 2013). A petition calling on the UK government to 'find a durable solution' for 'Somali M[oney] S[ervice] B[usinesse]s which provide financial lifeline services to Somalia and other developing countries' received over a 100,000 signatures.[13] The account closures were also debated in UK Parliament, with Rushanara Ali (MP for Bethnal Green & Bow) stating she had 'deep concerns about decisions that have been made in the past, not just by Barclays but by other banks such as HSBC, to remove banking facilities that are affordable for hard-pressed families who are trying to get support to other parts of the world' (quoted in Holley 2013a). One of the largest companies affected, a remittance network called Dahabshiil, contested Barclays' decision in court, on the grounds that it affected fair competition by driving selected money remitters out of the market. In November 2013, Dahabshiil won an interim judgement by the UK High Court which prohibited Barclays from closing the accounts until the case was decided at trial (Holley 2013b).[14] However, the case eventually did not go to trial because the companies reached a settlement in 2014, allowing Dahabshiil time to find alternative banking arrangements 'while working with the British government on a project to track payments made from the UK to Somalia' (Colchester 2014). If the case had gone to a full trial, Barclays would have been compelled to document its terrorism financing risk assessment practices and the ways in which it decides on risk appetite publicly. In addition, it would have been revealing to see a court weigh Barclay's obligations in counter-terrorism finance compliance against the question of whether banking is a social utility or a right. During the preliminary case between Barclays and Dahabshiil, the court acknowledged that 'there is no formal banking system in Somalia' and that there are 'clear social and humanitarian benefits associated with an efficient and properly regulated system for the transmission of money to Somalia'.[15] Decoupling these money service businesses from banking services thus affects what Zelizer calls the 'connected lives' that are built, maintained and negotiated through remittance networks (2005, p. 32; also Cooper and Walker 2016).

The account closures have fuelled an international debate on what has come to be called 'derisking', understood by the FATF as 'the phenomenon of ... restricting business relationships with ... categories of clients to avoid, rather than manage, risk'.[16] In an environment where, as Barclays put it, it is not possible to 'spot criminal activity with the degree of confidence required',[17] but where public association with terrorism financing could do very serious damage to a company's reputation and be grounds for hefty OFAC fines, banks apparently deem it better to preemptively exit these risky sectors altogether. According to Tom Keatinge, derisking decisions are based new types of 'unquantifiable risk' related to money laundering and terrorism financing fines, and potential 'worst-case scenarios' about 'what would happen if ... [a bank's] dealings with this client were to end up on the front page of the newspaper' (Keatinge 2014, pp. 50–1; cf. Power 2005). The broad and unspecified concept of risk appetite functions as a quasi-juridical term that suggests a process of solid risk assessment, but which in fact functions to legitimate preemptive decisions under conditions of uncertainty. Here, business objectives and commercial bottom lines become grafted onto the security functions of anti-terrorism financing screening. In fact, as became clear during the Dahabshiil case, Barclays introduced new financial thresholds for MSB customers in its risk assessment, and judged that 'it would not be commercially viable to continue to provide services to any customer representing less than £100,000 in annual revenue'.[18] FATF for its part stresses that denying banking facilities to entire business sectors is not what the risk-based approach to terrorism financing regulation intends. Even if there is no direct causal relation between the HSBC 2012 settlement and the Barclays 2013 decision to unbank Dahabshiil, there is certainly a relation of what William Connolly (2005, p. 870) calls 'resonance' in which 'heretofore unconnected or loosely associated elements fold, bend, blend, emulsify, and dissolve into each other, forging a qualitative assemblage'. It has to be recognized, furthermore, that derisking affects client groups very unevenly. While European banks continue relationships with the contested Saudi bank al-Rahji, Somali remittance companies and Islamic faith-based charities have faced numerous account closures.

Conclusion

In the wake of the Paris attacks of November 2015, the G20 released a statement reaffirming the commitment to 'tackling the financing channels of terrorism' and implementing FATF standards in 'all jurisdictions'.[19] The pursuit of suspect monies and potential terrorism financing is a seemingly targeted and relatively non-violent response in the war on terror, that places banks in the frontline of global security practice. This chapter has examined the increasingly pressing agenda of pursuing terrorism financing by starting with the US$1.9 billion settlement between HSBC and OFAC, and argued that it entails important and relatively invisible reconfigurations of geographical jurisdiction and financial client risk. In its investigation of HSBC, the US Senate offered an understanding of its jurisdiction as transactional rather than territorial, with important implications for *all* US-dollar denominated transactions. This move takes place in the context of European contestations of the legitimacy of sanctions lists on the basis of human rights concerns. The US Senate's investigation and settlement functioned to enforce OFAC sanctions lists extraterritorially, and instilled a risk-averse attitude in European banks' compliance departments. I have argued that the OFAC fines – not just in the case of HSBC, but also as directed against other European banks including Bank Paribas and Deutsche Bank – have resulted in new risk assessments and client risk profiles, for example concerning the MBS sector. In some cases, banks now prefer to terminate rather than monitor high-risk client relationships, especially in the case of low-profit but high-risk sectors like money transfer services and charities.

Conceptually, I have offered the notion of a finance-security assemblage to capture the particular transnational configurations of power in the pursuit of terrorism financing. Here, traditional national jurisdictional boundaries are circumvented and US sanctions decisions become enforced with global reach. My analysis has suggested that a finance/security assemblage exerts substantial power through administrative fines, whereby the US Senate investigation resonates with technical compliance decisions and banks' risk appetite assessments, resulting in the debanking of vulnerable client groups, including remittance-dependent families. These configurations work through little-known administrative procedures, via commercial risk assessment modules, to affect the everyday lived realities of migrant networks. Understanding these developments as an assemblage is not meant to downplay the hegemonic power of the US in this domain. Instead, it '*broadens* the range of places to look for... the sources of harmful effects' and looks to 'long-term strings of events' (Bennett 2010, p. 37, emphasis added). It takes seriously the independent decisions inside banks and the power of technical risk assessments, and draws attention to multiple potential sites of responsibility for harmful effects including derisking.

Acknowledgements

Many thanks to the editors, Brett Christophers, Andrew Leyshon and Geoff Mann for inviting me to be part of this project, and for their generous editorial comments. Financial support for this research is provided through the European Research Council (ERC-2015-CoG 682317 – FOLLOW).

Notes

1 Quoted in French Government, Press Kit, *Countering Terrorist Financing*, November 2015, p. 4, available at: http://www.gouvernement.fr/sites/default/files/locale/piece-jointe/2015/12/countering_terrorist_financing.pdf (accessed 24 January 2016).

2 International Convention for the Suppression of the Financing of Terrorism, Adopted by the General Assembly of the United Nations in resolution 54/109 of 9 December 1999, available at: http://www.un.org/law/cod/finterr.htm (last accessed 2 December 2015).

3 Worldbank Press Release, *Remittances to Developing Countries Resilient in the Recent Crisis*, 8 November 2010, available at: http://web.worldbank.org/WBSITE/EXTERNAL/NEWS/0,,contentMDK:22757744 ~ page-PK:64257043 ~ piPK:437376 ~ theSitePK:4607,00.html (last accessed 2 December 2015).

4 http://www.justice.gov/opa/speech/assistant-attorney-general-lanny-breuer-speaks-hsbc-press-conference.

5 http://www.treasury.gov/resource-center/sanctions/CivPen/Pages/civpen-index2.aspx

6 http://www.treasury.gov/resource-center/sanctions/SDN-List/Pages/default.aspx.

7 See Press Release (6 September 2013) at: http://curia.europa.eu/jcms/upload/docs/application/pdf/2013-09/cp130099en.pdf.

8 Available here: http://www.wsj.com/articles/SB10981352187914849 2.

9 Dahabshiil Transfer Services Ltd v Barclays Bank Plc [2013] EWHC 3379 (2013), p. 10.

10 Dahabshiil Transfer Services Ltd v Barclays Bank Plc [2013] EWHC 3379 (2013), p. 6.

11 Barclays statement in response to Change.org decision, available at: https://www.change.org/p/number10gov-stop-moving-the-goal-posts-and-take-every-step-you-can-to-ensure-remittances-flow-through-safer-channels-to-somalia/responses/9222 (last accessed 30 November 2015).

12 Barclays statement in response to Change.org decision, available at: https://
www.change.org/p/number10gov-stop-moving-the-goal-posts-and-take-
every-step-you-can-to-ensure-remittances-flow-through-safer-channels-to-
somalia/responses/9222 (last accessed 30 November 2015).
13 Available at: https://www.change.org/p/number10gov-stop-moving-the-goal-
posts-and-take-every-step-you-can-to-ensure-remittances-flow-through-
safer-channels-to-somalia (last accessed 30 November 2015).
14 Dahabshiil Transfer Services Ltd v Barclays Bank Plc [2013] EWHC 3379
(2013).
15 Dahabshiil Transfer Services Ltd v Barclays Bank Plc [2013] EWHC 3379
(2013), pp. 7–9.
16 As discussed during the October 2014 FATF Plenary Meeting, see: http://
www.fatf-gafi.org/documents/news/rba-and-de-risking.html (last accessed
30 November 2015).
17 Barclays statement in response to Change.org decision, available at: https://
www.change.org/p/number10gov-stop-moving-the-goal-posts-and-take-
every-step-you-can-to-ensure-remittances-flow-through-safer-channels-to-
somalia/responses/9222 (last accessed 30 November 2015).
18 Dahabshiil Transfer Services Ltd v Barclays Bank Plc [2013] EWHC 3379
(2013), p. 6.
19 Statement can be accessed at: http://www.consilium.europa.eu/en/press/
press-releases/2015/11/16-g20-leaders-antalya-statement-terrorism/(last
accessed 30 November 2015).

References

Acuto, M., and Simon C. 2013. *Reassembling International Theory: Assemblage Thinking and International Relations*. Palgrave Pivot. Available at: http://link.springer.com/book/10.1057%2F9781137383969 (accessed: 20 February 2017).
Agamben, G. 2009. *What is an Apparatus? And Other Essays*. Sanford, CA: Stanford University Press.
Aitken, R. 2006. Capital at its fringes. *New Political Economy* 11 (4):479–96.
Allen, J. 2011. Powerful assemblages? *Area* 43 (2):124–7.
Allen, J. and C. Allan. 2007. Beyond the territorial fix: Regional assemblages, politics and power. *Regional Studies* 41 (9):1161–75.
Amicelle, A., and J. Elida. 2016. The cross-colonization of finance and security through lists: banking policing in the UK and India. *Environment and Planning D: Society and Space* 34 (1):89–106.
Amoore, L. 2009. Lines of sight: On the visualization of unknown futures. *Citizenship Studies* 13 (1):17–30.
Amoore, L. 2011. Data derivatives: On the emergence of a security risk calculus for our time. *Theory, Culture & Society* 28 (6):24–43.
Amoore, L. and M. de Goede. 2008. Transactions after 9/11. *Transactions of the Institute of British Geographers* 33 (2):173–85.
Anderson, B., and C. McFarlane. 2011. Assemblage and geography. *Area* 43 (2):124–227.

Aradau, C., and R. van Munster. 2007. Governing terrorism through risk: Taking precautions, (un)knowing the future. *European Journal of International Relations* 13 (1):89–115.

Arnoldi, J. 2000. Derivatives: virtual values and real risks. *Theory, Culture & Society* 21 (6): 23–42.

Bagley, D. 2012. Written testimony for senate permanent subcommittee on investigations. Washington, 17 July, available at: http://www.hsgac.senate.gov/subcommittees/investigations/hearings/us-vulnerabilities-to-money-laundering-drugs-and-terrorist-financing-hsbc-case-history (accessed 10 December 2015).

Barrett, D. 2014. Muslim banks accounts closed by HSBC in wake of 'money laundering' fine. *Telegraph*, online edition, available at: http://www.telegraph.co.uk/news/uknews/terrorism-in-the-uk/11000794/Muslim-bank-accounts-closed-by-HSBC-in-wake-of-money-laundering-fine.html (accessed 10 December 2015).

Bennett, J. 2005. The agency of assemblages and the North American blackout. *Public Culture* 17 (3):445–65.

Bennett, J. 2010. *Vibrant Matter: A Political Ecology of Things*. Durham NC: Duke University Press.

Bialasiewicz, L., D. Campbell, S. Elden, S. Graham, A. Jeffrey and A.J. Williams. 2007. Performing security: The imaginative geographies of current US strategy. *Political Geography* 26:405–22.

Biersteker, T.J. 2004. Counter-terrorism measures undertaken under UN Security Council auspices. In *Business and Security: Public–Private Relationships in a New Security Environment*, eds A.J.K. Bailes and I. Frommelt. Oxford: Oxford University Press.

Biersteker, T.J., and S.E. Eckert, eds. 2007. *Countering the Financing of Terrorism*. London: Routledge.

Boltanski, L., and L. Thévenot. 2000. The reality of moral expectations: a sociology of situated judgement. *Philosophical Explorations* 3 (3):208–31.

Brown, J.M. 2011. BAE agrees to buy Norkom for €271 m. *Financial Times*, 14 January, online edition, available at: http://www.ft.com/intl/cms/s/0/c785fd62-1fb0-11e0-b458-00144feab49a.html (accessed 10 December 2015).

Brown, G. 2006. Securing our future. *The RUSI Journal* 151 (2):12–20.

Bueger, C. 2013. Thinking assemblages methodologically. In *Reassembling International Theory: Assemblage Thinking and International Relations*, eds Michele Acutoe and Simon Curtin. Palgrave Pivot. Available at: http://link.springer.com/book/10.1057%2F9781137383969 (accessed: 20 February 2017).

Christophers, B. 2012. Book review speculative security: The politics of pursuing terrorist monies. *Society & Space*: http://societyandspace.com/reviews/reviews-archive/de-goede-marieke-2012-speculative-security-reviewed-by-brett-christophers/(accessed 10 December 2015).

Colchester, M. 2014. Barclays settles dispute with money transfer company Dahabshiil. *Wall Street Journal* 16 April, online edition, available at: http://www.wsj.com/articles/SB10001424052702304626304579505661804336976 (accessed 10 December 2015).

Connolly, W.E. 2005. The evangelical–capitalist resonance machine. *Political Theory* 33 (6):869–86.

Cooper, M. 2010. Turbulent worlds: financial markets and environmental crisis. *Theory, Culture, Society* 27 (2–3):167–90.

Cooper, K., and C. Walker. 2016. Security from terrorism financing: Models of delivery applied to informal value transfer systems. *British Journal of Criminology* 56 (6):1125–45.

Coward, M. 2009. Network-centric violence, critical infrastructure and the urbanization of security. *Security Dialogue* 40 (4–5):399–418.

De Goede, M. 2011. Blacklisting and the ban: Contesting targeted sanctions in Europe. *Security Dialogue* 42 (6):499–515.

De Goede, M. 2012. *Speculative Security: The Politics of Pursuing Terrorist Monies.*, Minneapolis: University of Minnesota Press.

De Goede, M., and S. Randalls. 2009. Preemption, precaution: Arts and technologies of the governable future. *Environment and Planning D: Society and Space* 27 (5):859–78.

De Goede, M., and G. Sullivan. 2016. The politics of security lists *Environment and Planning D: Society and Space* 34 (1):67–88.

Deloitte, 2011. *Final Study on the Application of the Anti-Money Laundering Directive*. European Commission, DG Internal Market.

Durner, T., and L. Shretret. 2015. *Understanding Bank Derisking and Its Effects on Financial Exclusion*, Global Center on Cooperative Security.

Elden, S. 2007. Terror and territory. *Antipode* 39 (5):821–45.

FATF. 2015. *Financing of the Terrorist Organisation ISIL*. Paris: February.

Favarel-Garrigues, G., T. Godefroy and P. Lascoumes. 2008. Sentinels in the banking industry: Private actors and the fight against money laundering in France. *British Journal of Sociology* 48 (1):1–19.

Foucault, M. 2007. *Security, Territory, Population: Lectures at the Collège de France 1977–1978*, ed. Michel Senellart, trans. Graham Burchell. Basingstoke: Palgrave Macmillan.

French, S., A. Leyshon and T. Wainwright. 2011. Financializing space, spacing financialization. *Progress in Human Geography* 35:798–819.

Germain, R.D. 1997. *The International Organisation of Credit: States and Global Finance in the World-Economy* Cambridge: Cambridge University Press.

Gilbert, E. 2013. Introduction: Reading *Speculative Security*. *Political Geography* 32:52–4.

Gilbert, E. 2015a. Money as a 'weapons system' and the entrepreneurial way of war. *Critical Military Studies* 1 (3):202–19.

Gilbert, E. 2015b. The gift of war: Cash, counterinsurgency and 'collateral damage'. *Security Dialogue* 46 (5):403–21.

Gregory, D. 2004. *The Colonial Present: Afghanistan, Palestine, Iraq*. Malden, MA: Blackwell.

Gregory, D., and A. Pred. 2007. Introduction. In *Violent Geographies: Fear, Terror and Political Violence*, eds Derek Gregory and Allan Pred. New York & London: Routledge.

Helleiner, E. 1994. *States and the Reemergence of Global Finance: From Bretton Woods to the 1990s*. Ithaca: Cornell University Press.

Heng, Y-K., and K. McDonagh. 2008. The other war on terror revealed: Global governmentality and the financial action task force campaign against terrorist financing. *Review of International Studies* 34 (3):553–73.

High Court of Justice UK. 2003. *Al Rajhi Banking and Investment Corporation versus Wall Street Journal Europe*, Queen's Bench Division, 21 July [2003] EWHC 1358 (QB), available at: http://www.bailii.org/cgi-bin/markup.cgi?doc=/ew/cases/EWHC/QB/2003/1358.html (accessed 10 December 2015).

Holley, E. 2013a. Barclays under fire for 'outrageous' remittance closures. *Banking Technology* 20 July, available at: http://www.bankingtech.com/154562/barclays-under-fire-for-outrageous-remittance-closures/.

Holley, E. 2013b. Barclays loses High Court battle on remittance closures. *Banking Technology* 5 November, available at; http://www.bankingtech.com/180791/barclays-loses-high-court-battle-on-remittance-closures/.

Holodny, E. 2015. 2015 Cculd be the year we witness the weaponisation of finance. *Business Insider*, 5 January, available at: http://uk.businessinsider.com/weaponization-of-finance-eurasia-group-2015-1?r=US&IR=T.

Howell, J., and J. Lind. 2009. *Counter-Terrorism, Aid and Civil Society: Before and After the War on Terror*. Basingstoke: Palgrave Macmillan.

Hülsse, R., and D. Kerwer. 2007. Global standards in action: Insights from anti-money laundering regulation. *Organization* 14 (5):625–42.

Keatinge, T. 2014. *Uncharitable Behaviour*. London: Demos.

Langley, P. 2008. *The Everyday Life of Global Finance: Saving and Borrowing in Anglo-America*. Oxford: Oxford University Press.

Larner, W., and N. Laurie. 2010. Travelling technocrats, embodied knowledges: Globalizing privatisation in telecoms and water. *Geoforum*. 41:218–26.

Laurie, D. 2014. HSBC closes some Muslim groups' accounts. *BBC NEWS* 30 July: available at: http://www.bbc.com/news/business-28553921 (accessed 10 December 2015).

Lehto, M. 2009. *Indirect Responsibility for Terrorist Acts: Redefinition of the Concept of Terrorism Beyond Violent Acts*. Leiden, Brill.

Levi, M. 2010. Combating the financing of terrorism: A history and assessment of the control of 'threat finance'. *British Journal of Criminology* 50 (4):650–96.

Leyshon, A. 1997. Geographies of money and finance II. *Progress in Human Geography* 21 (3):381–92.

Leyshon, A., and N. Thrift. 1999. Lists come alive: Electronic systems of knowledge and the rise of credit-scoring in retail banking. *Economy and Society* 28 (3):434–66.

Leyshon, A., and N. Thrift. 2007. The capitalization of almost everything: The future of finance and capitalism. *Theory, Culture & Society* 24 (7–8):97–115.

Li, T.M. 2007. Practices of assemblage and community forest management. *Economy & Society* 36 (2):263–93.

Malkin, L., and Y. Elizur. 2002. Terrorism's money trail. *World Policy Journal*. Spring:60–70.

Maurer, B. 2005. Due diligence and 'reasonable man' offshore. *Cultural Anthropology* 20 (4):474–505.

Merton, R.C. 1998. Application of option-pricing theory: Twenty-five years later. *American Economic Review* 88 (3):323–49.

Muir, H. 2013. Mo Farah pleads with Barclays not to end remittances to Somalia. *Guardian*, 26 July, online edition: http://www.theguardian.com/sport/2013/jul/26/mo-farah-barclays-remittance-market (accessed 10 December 2015).

New York Times. 2003. Charity leader tied to terrorism gets prison term. *New York Times* 9 August: http://www.nytimes.com/2003/08/19/us/charity-leader-tied-to-terrorism-gets-prison-term.html?ref=topics.

Oborne, P. 2015. Why did HSBC shut down bank accounts? BBC News, 28 July, online edition, available at: http://www.bbc.com/news/magazine-33677946 (accessed 10 December 2015).

Opitz, S., and U. Tellmann. 2011. Global territories: Zones of economic and legal dis/connectivity. *Distinktion* 13 (3):261–82.

Opitz, S., and U. Tellmann. 2014. Future emergencies: Temporal politics in law and economy. *Theory, Culture & Society* 32 (2):107–29.

Power, M. 2005. The invention of operational risk. *Review of International Political Economy* 12 (4):577–99.

Pryke, M., and J. Allen. 2000. Monetized time–space: Derivatives – money's new imaginary? *Economy and Society* 29 (2):264–84.

Roberts, S. 1994. Fictitious capital, fictitious space: The geography of offshore financial flows. In *Money, Power, Space*, eds Stuart Corbridge, Ron Martin and Nigel Thrift. Oxford: Basil Blackwell.

Simpson, G.R. 2007. US tracks Saudi bank favored by extremists. *Wall Street Journal* 26 July, online edition, available at: http://www.wsj.com/articles/SB118530038250476405 (accessed 10 December 2015).

Sullivan, G. 2014. Transnational legal assemblages and global security law: Topologies and temporalities of the list. *Transnational Legal Theory* 5 (1):81–127.

Tickell, A. 2003. Cultures of money. In *Handbook of Cultural Geography*, eds Kay Anderson, Mona Domosh, Steve Pile and Nigel Thrift. London: Sage.

UK Treasury. 2007. *The Financial Challenge to Crime and Terrorism*. London, February.

US Senate. 2012a. *US Vulnerabilities to Money Laundering, Drugs, and Terrorist Financing: HSBC Case History*, Printed Hearing Record, Volume I, available at: https://www.gpo.gov/fdsys/pkg/CHRG-112shrg76061/pdf/CHRG-112shrg76061.pdf (accessed 10 December 2015).

US Senate. 2012b. *US Vulnerabilities to Money Laundering, Drugs, and Terrorist Financing: HSBC Case History*, Majority and Minority Staff Report, Permanent Subcommittee on Investigations, available via: http://www.hsgac.senate.gov/subcommittees/investigations/hearings/us-vulnerabilities-to-money-laundering-drugs-and-terrorist-financing-hsbc-case-history (accessed 10 December 2015).

Vlcek, W. 2008. A leviathan rejuvenated: Surveillance, money laundering and the war on terror. *International Journal of Politics, Culture and Society* 20 (1–4):21–40.

Warde, I. 2007. *The Price of Fear: Al-Qaeda and the Truth Behind the Financial War on Terror*. New York: I.B. Tauris.

Zarate, J.C. 2013. *Treasury's War*. New York: Public Affairs.

Zedner, L. 2007. Pre-crime and post-crimology? *Theoretical Criminology* 11 (2):261–81.

Zelizer, V. 2005. *The Purchase of Intimacy*. Princeton NJ: Princeton University Press.

6

Undoing Apartheid?

From Land Reform to Credit Reform in South Africa

Deborah James

Undoing Apartheid? From Land Reform to Credit Reform in South Africa

Forms of injustice left over when an authoritarian regime is in its dying throes, says legal anthropologist Sally Falk Moore, are never easy to eradicate. At moments of fundamental political transition, there are limits on the ability of new governments, and the new laws they pass, to shape the activities and dispositions of citizens. Times of change, often preceded by struggle and suffering, seem to offer the opportunity to rearrange political and economic regimes and to deliver justice to those who never before dared to hope for better, but implementing 'reversionary legislation' is difficult (Falk Moore 2011). What happens to those with an existing stake in the system? How far can new arrangements be forced to accommodate the continuation of certain older ones? When governments use wide-reaching legislative programmes to implant new regimes, their decrees sometimes simply exist alongside local mores: existing 'ways of behaving … exist parallel to and under the eyes of a new government' (ibid: p. 7). Alterations can seldom be forced by fiat. This paper[1] looks through the prism of land reform, which attempted to address one of apartheid's more notorious and obvious aspects of dispossession, to explore some of these questions about the limits to planned and legislated intervention in a related, less obvious, and lesser-known aspect of South Africa's discriminatory system: 'credit apartheid'.

Money and Finance After the Crisis: Critical Thinking for Uncertain Times,
First Edition. Edited by Brett Christophers, Andrew Leyshon and Geoff Mann.
© 2017 John Wiley & Sons Ltd. Published 2017 by John Wiley & Sons Ltd.

Lack of access to both land and credit was key in underpinning apartheid. Both aspects proved to need reform when, more than twenty years back, South Africa gained its political freedom. At the moment of democracy in 1994, returning land to those deprived of it appeared the most pressing. Credit reform, although already perceived as necessary, was not yet as urgent. When that urgency increased, around ten years later, it was partly because of what had happened in the meantime. In part reflecting the spread of financialization worldwide, but also for reasons specific to the local context, credit provision and take-up had rapidly escalated over a very short period. New sources were added to existing ones as access to credit was democratized and black people demanded, and obtained, access to credit at previously unheard-of levels. The rapid growth of a new black middle class during the 1990s would have been impossible without borrowing, but the debts newly incurred by this group, as by those aspiring to similar levels of upward mobility, were proving unserviceable.

Land reform did not materialize as it was supposed to. More longed for and planned for than achieved, it proved to be (and has increasingly become) a smokescreen for other more urgent deliverables which might have fulfilled the promises of citizenship more effectively. Seemingly more unattainable as the 1990s wore on, these included regular and well-paid employment for all, and the efficient provision of municipal services. Many of land reform's intended beneficiaries were in any case more oriented to modern conditions and wage work or salaried employment than to 'back-to-the-land' rurality. So the promises of democracy had to be delivered by other means. Although a number of the newly aspirant were employed, especially in public service jobs, the requirements of their new middle-class lifestyle outweighed the relatively modest salaries they were paid. The only way to get a house, a car, send one's children to university – as well as supporting less fortunate relatives – was on credit.

Thus, at precisely the moment when surging levels of consumer indebtedness made the need to regulate credit access most visible and urgent, the putative beneficiaries of such reforms had most to lose from regulation. This has proved so far to be an insurmountable obstacle. Because credit was having to carry the whole weight of the expectations and aspirations democracy unleashed – those expectations upon which the somewhat misguided initiatives of land reform had *not* delivered – attempts to regulate it were strenuously resisted, not only by the new lenders who were cashing in on profits from high-interest unsecured loans, but also by their clients who wanted to carry on borrowing.

In sum, the short period of twenty-something years since the advent of democracy has seen a rapid transition. An earlier era characterized by evocative discourses about restored citizenship, to which land was a concrete means, has been brought into sudden and uncomfortable juxtaposition with a later era dominated by neoliberal-style speculation. Alongside, and gradually supplanting, a discourse of land as ensuring livelihoods and well-being for the not-so-well-off, new enunciations emphasize land's potential as an 'asset class' for would-be investors. The transition from a state-planned economy to a rapidly liberalizing one has seen these older discourses disappear, introducing newer ones as if by stealth. Where land is concerned, what was previously seen as a means to regain sovereignty is now, where money is concerned, something in which financiers have begun to show interest via sovereign wealth funds and the like (Ducastel and Anseeuw 2013). The original land dispossession through which black people were turned into wage slaves has morphed into a new dispossession characterized by the characteristically twenty-first-century form of debt slavery: 'working for *mashonisa*' (the loan shark). Paradoxically, this developed to its fullest extent only after the moment of political freedom. Whereas the deprivations involved in land alienation were enacted upon an unwilling populace, the disadvantages of overindebtedness are more insidious because they entail willing participation and collusion on the part of those involved.

In the case of both land and credit, older practices and institutions have proven able to adapt, existing in parallel with a new regime. Explaining the limits to reform lies beyond invoking a nation's reluctance or inability to police and enforce the new laws (once in place). It goes beyond the assertion that the old laws might have worked perfectly well had they been properly enforced – although that is certainly part of it. It also requires more than simply asserting that established interests resist transformation. (Critical accounts have shown that the South African democratic transition was limited in scope (Adam et al. 1998; Bond 2000; Marais 2001) because the struggle leaders-turned-liberation-elite struck deals with the corporate sector which meant that there were very few changes in the character of South African capitalism.) Though that, too, is part of the explanation.

Further reasons for the peculiarity, and partial failure, of this reversionary legislation must be sought at other levels. Before exploring these in detail, the first section of the paper gives a brief account of the recent lending boom and the inability of existing legal frameworks to keep it in check. The second explains the *longue durée* of both land and credit apartheid in historical terms. The third gives an account of the attempts, post-apartheid, to impose new regulations. Among several reasons these failed to take root, as the fourth section demonstrates, is the fact that, given the longstanding character of these twin systems of discrimination, numerous intermediaries and agents have entrenched interests in the status quo. The very attempts to tackle this problem by incorporating the marginal and previously politically disenfranchised, and to create a single economic framework from a dual one by changing the character of savings, consumption, investment and property ownership, have simply ensconced these interests still further. All of these factors play key roles in shaping a peculiar contradiction: a perceived need to restrict a credit system for the good of people who, by and large, would prefer that borrowing and lending continue unchecked.

Credit Boom/South Africa's Financialization

Credit apartheid, in its newest incarnation, is a localized, vernacularized version of the global phenomenon we have come to know as financialization: something said to have taken hold in South Africa in the late 1990s and early 2000s, accounting for the country's 'jobless growth' during that period (Barchiesi 2011; Marais 2011, pp. 124–8, 132–9). Here, as elsewhere in the globe, this recently evolving version of capitalism, with a growing preponderance of finance in economy, politics and everyday life, has not only transformed society as a whole (Friedman and Friedman 2008; Kalb 2012, 2013), but also implanted itself in the everyday of communities and households via the 'financialisation of daily life' (Langley 2008; Martin 2002; Krige 2014). From creditors' point of view, financialization describes a new 'pattern of accumulation in which profit making occurs increasingly through financial channels rather than through trade and commodity production'; for borrowers, it means they are 'confronted daily with new financial products' (Krippner 2005, pp. 173–4), and those previously unschooled in matters of saving, borrowing and repaying are enjoined actively to start modelling their use of money along more formal lines, and to become 'financially literate' (Kear 2013).

In South Africa, data show a huge growth in borrowing during the early 1990s. In 2011, household debt stood at R1.2-trillion, up from R300-billion in 2002: as a percentage of disposable income it was 76 per cent, ratcheting up from 50 per cent in 2002. In 2014, a World Bank report claimed that South Africans were borrowing more than any other country.[2] The particularities of the South African transition help to explain this apparent epidemic of borrowing. Following the first democratic elections in 1994, a concerted attempt to abolish the various aspects of apartheid coincided with a massive rise in expectations. This demand for credit was met with a burgeoning supply. As members of a rising black middle class replaced the (mostly white) incumbents of the previous civil service, these newly-redundant public servants started lending money to those replacing them – and others – at high rates of interest. Other micro-lenders soon joined them, some formal and technically legal, others informal and classified as loan sharks. Big banks and mainstream retailers, many of which had formerly been reluctant to offer credit to black people, joined in. Salaries deposited in people's bank accounts thus started circulating throughout the system, with a considerable proportion – plus interest – ending up in lenders' pockets. In the case of salaried and waged borrowers, lenders incurred few risks: they were using borrowers' salaries as a form of collateral (Roth 2004, 78; see Anders 2009, p. 76; Maurer 2012). But in those cases termed 'unsecured lending', where collateral was lacking, interest rates were even higher. This sharp rise in interest rates was made possible when, during the period of rapid liberalization in the 1990s, the state repealed the terms of the Usury Act in the interests of expanded credit access. The Reserve Bank reported that unsecured lending had nearly tripled between 2009 and 2013 to 10.5 per cent of total credit. Borrowers in the unsecured category were *not* repaying their loans: nearly half were at least three months behind on debt payments.[3]

For decades before this, however, borrowers did keep up their payments, at least enough to sustain the business model that underpinned credit apartheid. In this highly exclusionary system, money was lent to those in disadvantaged groups, but only in unequal ways. People disallowed from land ownership, as the next section explains, had few credit options. Buying movable property – furniture or appliances – was the next best thing to real estate. The initial purchase often accompanied a daughter's marriage and served as part of her trousseau, after which she and/or her husband would buy further items on instalments, usually at more than twice the cash price, and pay them off item by item. These were long standing arrangements: 'on all sides', observed American social worker Ray Phillips in Johannesburg in the 1930s, 'the "hire-purchase" system of acquiring pianos and furniture is responsible for much of the indebtedness of the Africans' (1938, p. 41). The system was somewhat abused by clients inasmuch as they often bought items from a variety of merchants and were unable to pay them all – 'The ones who come first get the money, the rest are put off with various promises until a later time' (Phillips 1938, p. 40) – but it was also laid at the retailers' door for failing to check their clients' ability to pay. 'They do not care who comes into the shop, whether he or she earns £2 10s. 0d. per month, they give him or her goods worth £40 or £50, payable monthly at 30s. or 40s. per month. Then the buyer ... goes to another shop where he contracts another debt and so on continually.' In the long term, these retail businesses proved profitable: by the 2000s their practices were singled out by reformers as laying the basis for credit apartheid (Department of Trade and Industry [DTI] 2002, 2004).

During fieldwork in 2008, I got some insights into arrangements that had evolved to 'catch' the defaulters from Xolela May, a consumer rights activist and lawyer. Growing up in the black township of Langa, he regularly saw his neighbours approached by the sheriff of the court to arrange an inventory of their possessions prior to confiscation, while they stood by helplessly. Retailers were no longer willing to be 'put off with ... promises', as in the 1930s. By the time Xolela had studied law and joined the human rights organization the Black Sash in the 1990s, a far wider variety of lenders had begun extending credit to people formerly denied it – and this time repayments were easier to secure. Debtors were taken to court in record numbers, and repossessions carried out to an even greater extent than in Xolela's youth. Creditors, no longer standing in line to get their money, were now able to take it directly from debtors' salaries or wages, using 'garnishee' or 'emoluments attachment orders' (known as debt recovery orders in the UK).[4]

Such orders, requested by a creditor from a magistrate, require an employer to enable monthly repayment directly from the salary of a defaulting employee/debtor, with the creditor bearing a 5 per cent charge. The order, once granted, is served on the employer by a sheriff, but the debtor must agree by signing the order. It had become commonplace for employees – especially civil servants, but also train drivers, factory workers and supermarket employees – to have substantial parts of their salaries reclaimed by creditors before they were able even to see the money. 'Of 1.2 million to 1.3 million public servants at the national and provincial level, 210,000 were affected by garnishee orders', one newspaper reported in 2009.[5] Employers, noting the negative consequences for the well-being of workers – both civil servants and others – have taken some action. BMW, with its large car factories in the Eastern Cape, was granted funds by the German Development Funding Agency GTZ to commission detailed research on the effects on worker well-being of these debt recovery arrangements. It established that many practices used by debt collectors to 'attach'[6] workers' salaries were and are illegal. One irregularity concerned the use of signatures. Although the debtor must sign the 'consent to judgment,' as proof he or she has agreed to the arrangements, as the researcher Frans Haupt told me, 'if you have a legally and financially illiterate consumer he will sign anything, especially if you harass him at work'.[7] Another involved creditors, especially furniture retailers, obtaining orders from courts far from employees' places of work. A third involved debt collectors, paid on commission, forging debtors' signatures. (These collectors, and debt administrators who oversaw the process, were taking a cut at every stage of the process.) A fourth concerned lack of judicial oversight: frequently a relatively uneducated clerk of the court, rather than the magistrate, authorized these orders.[8]

Other creditors had more direct access to borrowers' bank accounts: unregistered loan sharks or *mashonisas*, lending to salaried or waged employees, would with borrowers' agreement keep their ATM cards and withdraw money owed to them at month end.[9] The system overall was skewed to the advantage of lenders, who had ways of getting their money back against all the odds, and who rarely lost out or received a sanction for failing to check clients' capacity to repay. The negative consequences ranged from people resigning from their jobs to escape creditors, through cashing in their pensions, to tragic measures such as suicide. The infamous shooting of striking miners at Marikana in 2012 was yet another manifestation of the problem. Newspaper reports revealed that the miners, not necessarily in the lowest pay bracket, carried unsustainable levels of debt. This was doubly burdensome, indeed intolerable, because of the way their numerous creditors ensured repayment. Miner's pay, automatically transferred into their bank accounts at month end, was immediately 'garnished' or transferred out again, so that shortly after payday, many of them had nothing left to live on.

All-in-all, alongside lenders' lack of attention to borrowers' credit-worthiness, illegal and unregulated collection practices set the scene for a continuation of credit apartheid: a system that arose out of systematic exclusion from certain lending arrangements, and the institutionaliza-tion – and almost inescapable inclusion – in others. Associated with an array of techniques of repossession, reckoning, evasion and eventually financialization, together with an increased demand for consumer goods and other signs of a good life, and a growing discrepancy between income and outgoings, this system had extraordinary tenacity, and seemed as impervious to regulation as other aspects of South Africa's segregationist arrangements, such as unequal access to land. To outline the relationship between these two modalities of apartheid before dis-cussing the attempted reforms, I draw on secondary sources from South Africa's rich radical history tradition.

Credit, land and labour: the *longue durée*

Land ownership and credit are self-evidently linked; conversely, the absence of landed property poses problems for lenders. The aspect of apartheid most notorious for its exclusionary character was legislated land dispossession. Black people moved, or were forcibly relocated, from areas the apartheid state designated as white farm land to the reserves (later bantustans or homelands) where they were subjected to imposed communal property arrangements. Many, being migrant labourers on the mines or in industry, had places to stay in the urban townships – either municipal hostel accommodation or state-owned housing – while others, settling in town more permanently, likewise rented their houses from the township authorities. In both cases they were largely excluded from owning property. In the absence of real estate to serve as collateral, the form of lending that came to predominate, as cultivators-become-migrants started buying furniture and appliances and including these in their customary marriage exchanges, and as retailers expanded their operations and aggressively marketed their products in the captive markets of the black homelands and townships, was hire purchase. The form of 'loan security' did not differ substantially: appliances were repossessed just as a house might have been. This was complemented and partly supplanted in the 1990s, when the new micro-lenders used civil service salaries paid directly into bank accounts 'as a collateral substitute' (Roth 2004, p. 78; Maurer 2012), or offered 'unsecured' loans at much higher rates of interest (Roth 2004, p. 41; Schoombie 2009), which they justified as a means of offsetting risk.

In South Africa's migrant economy, credit was linked not only to land (or the lack of it) but also to labour. To understand the particular form these linkages took, it is useful to contrast this with the situation that obtained in South East Asian and South Asian countries. There, paternalistic dependence combined with exploitation connected tenants to landlords who doubled as creditors, and who often milked their borrower/tenants dry over several generations while forcing them to continue providing labour (Murray Li 2010, Martin 2010; Mosse 2004; Shah 2010). South Africa's racial laws attempted a more definitive land dispossession. This was not easily implemented, however, and in many cases it was imperceptible changes in the economy of farming rather than draconian state measures that caused cultivators, eventually, to move off their lands. Although in some cases black people did, then, remain as tenants on land they did not own (Beinart and Delius 2014), they borrowed more often from trading-store owners than from white farmers: the resulting debt relations thus linked traders and their customers rather than landlords and their tenants. And the labour supplied as an offshoot of these arrangements was to mining and industry rather than to semi-feudal landlords.

Store owners, imbued with a free-wheeling and pioneering spirit of enterprise in the late nineteenth/early twentieth century, became the harbingers of commodification and market penetration in the rural areas where many black cultivators lived – although they plied their trade in eddies separate from the main stream of economic life, given that capital accumulation, following the mineral discoveries of the 1880s, centred on mining and farming. (Indeed, the racial/ethnic profiles of some of these traders, in a setting where many occupations were ethnically restricted and defined, ensured their marginality from the domain of production (see Hann and Hart 2012).)[10] While some traders were 'sympathetic' to their black customers (Van Onselen 1996, p. 186), others, in addition to selling trade-store goods, acted as independent labour recruiters who attempted to cajole black cultivators to become wage workers in the mine economy. Men buying cattle or goods on credit were obliged to 'work it off' after the event by enrolling for contract work on the mines (Beinart 1979; Schapera 1947), or were induced to enter into contracts by wages paid in advance. The system was much abused: traders often extended such large advances that the borrower 'remained in debt even after having worked for several months', while borrowers often accepted several advances simultaneously and never repaid them (Schapera 1947). Recruited through this and other methods, the workforce expanded, 'large sums of money' were injected into the rural economy (Crush 1986, p. 33), and recruiters, competing with each other, variously offered larger advances or smaller ones on better terms (for a similar case in West Africa see Martino 2017).

The colonial government intervened with characteristic paternalism. While regulating unsustainable practices which might have led to the collapse of agents' enterprises, it also took the line that workers' money was best removed from their control. Rather than allowing migrants to get ready access to their earnings, the authorities devised a system of deferring them, fearing that cash received immediately would be diverted from household expenses and the payment of various colonial government taxes and levies (Schapera 1947, pp. 106–7; First 1983). In sum, a combination of free-wheeling enterprise on one hand, and regulatory intervention on the other, laid the basis for a system in which earnings – and (much later) bank accounts used to transfer them – are viewed both as fair game for plunder and as legitimately controlled and/ or regulated from the outside. A deeply-rooted system of 'external judicial control' over wage-earners' finances (Haupt et al. 2008, p. 51) has resulted. Garnishee orders are its most recent variant.

Not all black people were restricted to reserve areas, however. Those who would later become members of the emerging middle class made efforts to procure their own land in the late nineteenth and early twentieth centuries. Some obtained title in recognition of former occupancy, others formed syndicates and bought land via missionaries and other intermediaries. But considerable opportunities for land speculation existed, and many black owners either lost their land after having borrowed against it, or willingly sold it to pursue education and other modern investments (Trapido 1978; Murray 1978). Those titleholders who remained on their pockets of land found themselves disadvantaged vis-à-vis white farmers, who received state subsidies, support from state marketing boards, and borrowed money on favourable terms from both commercial banks and the government's Land Bank (Beinart and Delius 1986, pp. 29–30; Morrell 1986, pp. 379–80). Despite these measures, some remained owners of their farms – so-called 'black spots' because of their situation amidst white-owned farms – from which they were forcibly removed. It was these titleholders that became a prime focus of some of the reform measures described below.

Both in its 'land' and its 'credit' variants, apartheid as fully developed by the Afrikaner nationalist government after 1948 was only the final, state-sanctioned version of what had started decades earlier in the combined forces of market and state. Inextricably interlinked from the outset, both involved exclusion from certain kinds of properties, facilities and loans, accompanied by virtually inescapable dependence on others.

Reversionary Legislation: New or Old Laws?

The post-apartheid land reform programme, aimed at providing redress to those affected by apartheid's land laws, involved extensively planned and high-visibility – albeit poorly-funded (Walker 2000) – legal processes. Instigated soon after 1994, the intention was to overturn or remedy very obvious, state-sanctioned, forms of land alienation, like the forced expulsion of titleholders from farms they had purchased, and resettlement in far-off and inhospitable sites in the reserves or homelands. Under restitution (see Table 6.1), 'black spots' were returned to titleholders and people and equipment transported back to start anew. The focus was overtly violent acts perpetrated by the state, especially brutal forced removals, conducted by the army and police and implemented in the name of state ideology and its racial laws.

In addition to these forms of redress, land reform also aimed to redistribute land and/or give secure tenure to those rendered homeless by the less visible forces of capitalism and the market – some of them still residing on white farms but with no legal rights to remain (see Table 6.1). Besides compensating those resettled by the authorities in the name of the state's racial laws, land reform also redistributed land, or gave a place to live, to those who had lost their places of residence because of changes in the economy of farming. State and market forces were entwined when the programme was implemented: government funds and advice were made available, but established property rights were respected through the market-driven 'willing buyer/willing seller' model. The restorative outcomes turned out, however, to be muted. Land reform had been perhaps too focused on the overt disruptions caused by state brutality to take account of those less obvious displacements attributable

Table 6.1 Land reform legislation

Category	Date	Act	Intention
Restitution	1994	Restitution of Land Rights Act	To provide for the restitution of rights in land to persons or communities dispossessed of such rights after 19 June 1913 as a result of past racially discriminatory laws or practices. To establish a Commission on Restitution of Land Rights (CRLR) and a Land Claims Court
Restitution/ Redistribution/ Tenure Reform	1996	Communal Property Associations Act **(CPA)**	To enable groups to acquire, hold and manage property as agreed by members and using a written constitution
Tenure Reform	1996	Land Reform (Labour Tenants) Act	To safeguard the rights of labour tenants who had been remunerated for labour primarily by the right to occupy and use land
	1996	Interim Protection of Informal Land Rights Act **(IPILRA)**	To protect people with informal rights and interests from eviction in the short term, pending more comprehensive tenure legislation (i.e., CLRA)
	1997	Extension of Security of Tenure Act **(ESTA)**	To give farm occupants rights of occupation on private land. Establishes steps to be taken before eviction of such people can occur
	2004	Communal Land Rights Act **(CLRA)**	To provide for legal security of tenure by transferring communal land to communities and provide for its democratic administration by them

Source: www.info.gov.za/gazette/acts; Adams 2000.

to the power of capitalism and ultimately far less easy to put right (James 2007). Those parts of the programme which were implemented often laid bare and even exacerbated preexisting inequalities (Hay 2014, 2015; James 2007; Falk Moore 2011, p. 13), and exposed the difficulties of 'reversionary legislation' (Falk Moore 2011, p. 14).

In contrast to the more obvious case of land dispossession (and its necessary corollary, land reform), concern with debtors' rights took longer to surface. Credit issues were less obviously a matter for the state, more obviously something to do with the market. 'Discriminatory legislation', if it existed at all, was not a uniquely South African phenomenon. On the contrary, the Magistrates' Court Act of 1944 had been imported from the UK with few modifications. Credit regulation, unlike reforms to more immediately visible aspects of the old order like land dispossession, had, on the face of it, a less obvious relationship to apartheid. Given that the new consumerism and its fall-out affected those from all races, it was far from obvious that the new excesses of lending, and the 'overborrowing' that resulted, had anything to do with the previous racist order. The everyday processes through which loans are granted, risks assessed, interest rates set, and property repossessed or wages docked when such loans are not repaid or when installments are in arrears for items bought on credit: all appear mundane and unremarkable, as well as being generic, seeming to vary little from one setting to another.

It was belatedly realized, however, that 'credit apartheid' had given these processes a very particular character – resulting in the kind of borrowing scenario outlined in the first section of the paper. Where other reversionary legislation (such as that concerning land) was passed in the 1990s and early 2000s, with special ministries or commissions founded to oversee them, those forms of regulation protecting the consumer/borrower, although likewise debated in the 1990s, were passed only towards the end of the 2000s – since, indeed, many of the worst problems had developed in their fullest measure only towards the end of the 1990s. They were left to the Department of Trade and Industry (DTI) to monitor and implement (DTI 2002; 2004).

As was the case with land reform, actions taken to regulate credit did not emerge from nowhere, however. They were built on the longer-standing activities of civil society protagonists and activists. For example, debt activists like Xolela May, in organizations such as the Black Sash, had long been devoted to addressing the problems of easy credit, indebtedness and lack of debtor protection that had become evident well before the end of apartheid. He helped develop the National Credit Act (NCA), effective from 2007 (see Table 6.2), along with its new system of debt review and debt counselling. Although a new public institution of sorts was established – the National Credit Regulator – its description as 'a consumer

Table 6.2   Credit reform legislation: Aims of the National Credit Act of 2007

Promoting the development of a credit market that is accessible to all
    South Africans, and in particular to those who have historically been
    unable to access credit under sustainable market conditions;
Ensuring consistent treatment of different credit products and different
    credit providers;
Promoting responsibility in the credit market by:

i. encouraging responsible borrowing, avoidance of over-indebtedness
    and fulfilment of financial obligations by consumers; and
ii. discouraging reckless credit granting by credit providers and
    contractual default by consumers;

Promoting equity in the credit market by balancing the respective rights
    and responsibilities of credit providers and consumers;
Addressing and correcting imbalances in negotiating power between
    consumers and credit providers by:

i.   providing consumers with education about credit and consumer rights;
ii. providing consumers with adequate disclosure of standardized
     information in order to make informed choices; and
iii. providing consumers with protection from deception, and from unfair
     or fraudulent conduct by credit providers and credit bureaux;

Improving consumer credit information and reporting and regulation of
    credit bureaux;
Addressing and preventing over-indebtedness of consumers, and
    providing mechanisms for resolving over-indebtedness based on the
    principle of satisfaction by the consumer of all responsible financial
    obligations;
Providing for a consistent and accessible system of consensual resolution
    of disputes arising from credit agreements; and
Providing for a consistent and harmonized system of debt restructuring,
    enforcement and judgement, which places priority on the eventual
    satisfaction of all responsible consumer obligations under credit
    agreements.

Source: based on http://www.justice.gov.za/mc/vnbp/act2005-034.pdf.

advocate that is charged with registering lenders' accurately depicts it as a
body with limited powers, not least because of disputes between different
government departments.[11]

Beyond the differing extents to which reform in these contrasting
cases relied on special bodies and government commissions, there are
further important contrasts. The strongly 'rights-oriented' character of
land reform legislation owed itself to the many lawyers employed in the
programme who had previously been human rights activists. It was a

result of – and a reaction to – the fact that apartheid South Africa had been 'quite self-consciously a legal order' in which 'nothing was done without legal authorization, from removals to detentions' (Chanock, quoted in Palmer 2001). By the same token, these activists-turned-new-government-officials never doubted that an entirely new swathe of laws would be required to facilitate reform. Like those responsible for the initial dispossession, those championing the dispossessed were preoccupied with the law and with legal rights in principle enforceable in court – and were perhaps overly optimistic concerning the law's capacity to facilitate planned social change (James 2007).

In the case of credit reform, although the need for new legislation belatedly became clear, some suggested that the problems might simply be addressed under existing legislation. Xolela May told me, for example, that legal practitioners motivated by profit to act on behalf of creditors were routinely ignoring particular sections of the relevant legislation, the Magistrates' Court Act of 1944, which might have afforded debtors some protection. One section of the Act required creditors to enquire into the financial position of debtors subject to garnishee or emoluments attachment orders. It provided debtors with greater rights than normally recognized, requiring that before garnishing part of the debtor's wages, 'the court has to ensure that that debtor has remained with sufficient means in order to maintain himself and his family'.[12] Ignorant of, or flagrantly ignoring, the spirit of the law, lawyers and the debt recovery companies for whom they worked were now intimidating debtors into agreeing to unsustainable repayments, which might amount to more than half their monthly income.

The Magistrates' Court Act, then, was not itself inattentive to the rights of consumers; rather, the problem was that it had not been properly enforced. Many of the abuses evident when reformers took a serious look at the nature of credit were not so much inherent in the law itself but rather in the way in which it had been implemented. One lawyer brought this up in the course of the hearings in parliament preceding the passage of the NCA. Rather than passing new legislation, he suggested, why not amend the Magistrates' Court Act by making it mandatory to have judgments 'dealt with in an open court of law' rather than, often fraudulently or in ignorance, by uneducated clerks of the court?[13] All in all, the fact that the legislation was implemented without due diligence is indicative of the continuing 'advantage to creditor' in South African law, which no new legislation has been able to address (Boraine and Roestoff 2002).

What of the key aim of credit reform – to combat the readiness with which lenders 'recklessly' extend loans to those patently unable to repay? Cases in which the court found against such lenders proved few and far between. In one instance, a home loan was extended to a person soon to retire, who would clearly have no means to repay it; in another, a number of credit providers, including several from the mainstream financial sector, were taken to court on behalf of debtors who had repayment commitments amounting to between 79 per cent and 160 per cent of their monthly income. The credit providers were found guilty. A third case demonstrates the determination of borrowers to obtain valued items at any cost: an officer in a bank was alleged to have been accepting bribes from customers in return for approving their loan applications; the bank was fined R305 million for reckless lending (James 2015, p. 89)

Some policymakers were hopeful they might have magistrates settle creditor/debtor disputes – and ultimately reinstate the cap on the interest rate – by law.[14] But magistrates, schooled in the old legislation, and too little acquainted with the new, proved unwilling or unable. Even in higher courts, where action might have been possible against the reckless excesses of credit capitalism, that capitalism's greater legal muscle held sway. Small successes by legal activists exploited areas of uncertainty between the old and the new forms of legislation, but no challenge proved sufficiently robust to qualify as reform. For the most part, then, it was the behaviour of borrowers rather than that of lenders reforms attempted to amend. A few refashioned citizens, a few small-time commission agents and debt collectors punished, seemed to be the most legislative efforts might yield.

Finally, the NCA had some unexpected outcomes, a description of which leads us to the last section of the chapter. The intention of the Act was not only to discipline 'reckless lenders' but also to provide a system more affordable to those owing relatively small amounts of money and for whom bankruptcy was unaffordable (Boraine and Roestoff 2002; Schraten 2014). It laid out measures aimed at stopping harassment by creditors, allowing debt rescheduling, and providing some breathing space in which to make payments while preventing further indebtedness. But such efforts at remedy and social justice were stymied, not only by debtors' determination to carry on borrowing, but also by the activities of intermediaries. In this case, just like the profiteering lawyers to whom Xolela alluded, existing debt administrators – a role specified under the old Magistrates' Court Act – had spotted an opportunity and had 'changed their hats' to become debt counselors – a role invented by the new legislation, the National Credit Act – seeing this as an alternative source of income. People at all levels now make money by charging commissions or adding interest at every point in the value chain. Over the *longue durée*, spaces have continually opened up for such figures, who play their role in establishing the current credit/debt landscape, and in resisting reversionary legislation.

Agents and Intermediaries

As hinted earlier, among the most important reasons why reform – of land or credit – was bound to be problematic owed itself to the role of intermediaries. Any apparent division between perpetrators and victims was inevitably blurred by the existence of actors with a wider range of entrenched interests than initially envisaged. Two brief examples illustrate this well.

Land reform turned out to be difficult to achieve. Despite being firmly regulated by the state, in the sphere of landholding and property ownership state law has long been out of kilter with everyday practice. When the state began its reform programme, it often found the wind taken out of its sails by small-scale intermediaries. Some were squatter landlords giving access to land not technically theirs to redistribute, while government planners, unaware of these machinations, hummed and hah-ed about how best to plan its use. Others were more savvy about the technical specificities of the law and the relevant bureaucracy, positioning themselves between community members and remote state or NGO officials. They developed followings among the rural underclass, promising the possibility of new land holdings. Such followings had the potential to make these intermediaries eligible, alongside their clientele, as land reform beneficiaries. Illegal or semiformal land selling or letting of this kind has been going on in different ways for over a century (Hay 2015; James 2011; for an earlier case see Stadler 1979).

Arguably, then, it is not so much that the laws before and since apartheid have been just or unjust in any one case, but that people have always had some ability to withstand these laws and simply get on with life. (Indeed, it took fifty years to enact the original discriminatory land laws, for much the same reason.) Overall, a simple binary between domination and resistance, conceived when legislation was designed, reduced 'the complexities of a historically produced politico-legal context' and obscured the existence of 'shifting patterns of dominance, resistance and acquiescence, which occur simultaneously' (Wilson 2001, cited in Falk Moore 2011, p. 9). The starkly oppositional image of a brutal state violating its people dissolved in the face of a more complex reality, with diverse groups, entrepreneurs and business interests colluding or resisting by turns, always defying easy classification into a schema of perpetrators and victims.

In the case of credit reform, various protagonists have taken up roles in the credit/debt machinery, in the process turning themselves into essential components of that machinery. Dodgy lawyers making money from debt recovery practices of borderline legality, and debt administrators-turned-debt-counsellors have already been mentioned. The hire purchase furniture business, because owners were socially and geographically remote from the bulk of their customers who lived in black townships and rural villages, relied – and continues to rely – on black agents. It has long been the case that the low pay and commission-based character of their work, combined with the exploitativeness of clients' borrowing terms, abetted the emergence of forms of illegal activity – 'scams' – in which agents acted in complicity with clients, enabling agents to augment their wages while clients temporarily evaded repayment. Echoing and modelling itself on this system, informal moneylending concerns also used agents, and these agents have similarly played the system while keeping their employers in the dark (James 2015, pp. 104–5, 113–15).

The inextricable collusion of people at all levels in making money on credit, if in a less patently 'illegal' manner, is demonstrated by a matter that came up when the NCA was debated as a bill in Parliament. The Act's attempt to tackle a broad spectrum of problems relating to 'reckless lending' included some which arose from the activities of the very brokers and agents outlined above. Its aims were ambiguous, however; it was intended not only to protect vulnerable and financially uninformed borrowers from unscrupulous creditors, but also, in the new spirit of affirmative action or 'Black Economic Empowerment', to open up new possibilities for black business in fields which had previously been dominated by whites or members of other ethnic minorities originating outside South Africa. Such opportunities included micro-lending, as well as occupations theoretically intended as part of the solution to indebtedness, such as debt administration and debt counselling. But the work such opportunities offered would itself require regulation and consumer protection. One such opportunity, a long-standing livelihood strategy in the black community, was 'direct selling' on credit (and on commission) at workplaces.

The rights of such sales agents were vigorously defended by one submission on the bill, which pointed out that the reason why they visit the 'work places of potential consumers to enter into loan agreements' was because such 'consumers are not able during office hours to attend at the credit provider's physical premises', and that worldwide trends in direct selling indicate that 'it is certainly convenient, speedy and efficient both for the credit provider and the consumer for the loan agreement at times, to be concluded at the consumer's work premises'. Prohibiting such a practice, the submission anticipated, would result in the closure or complete (and costly) restructuring of 'many small credit operator businesses relying solely on agents to sell their goods and/or products to employees at their work'.

The thousands of agents currently operating within the South African framework would immediately lose their jobs resulting in catastrophic implications for their families and extended families. ... by preventing business being done at work or at home a large section of the economy will effectively be destroyed overnight.[15]

The tone of the submission, made by Balboa, one of the new micro-lending organizations described in the first section of this chapter, seems to be attentive to the needs of small-scale entrepreneurs. Given that Balboa's business model depended on offering unsecured loans at high rates of interest the concern expressed here could be seen as veiling naked self-interest. Yet it is likely that many of the people who borrowed money from this and other micro-lenders, operating along the same lines as 'Avon Ladies' (Dolan and Scott 2009)[16] and similar multilevel marketers in schemes pervasive in South Africa, were using the borrowed money to establish their own enterprises. Borrowers, in other words, are also lenders (James 2015, pp. 8, 147, 198). The Act, in the end, did not require direct sellers to change their practices: to regulate might have been to circumscribe the activities of smaller operators and leave bigger ones untouched.

Overall, the case starkly illustrates the contradictory character of the legislation. Every piece of protection offered to borrowers might run the risk of forfeiting a semi-formal income-generating opportunity of the kind enthusiastically embraced by those whom credit apartheid formerly marginalized. Thus, while members of the business community reiterated the familiar claim that securing market freedom was in consumers' interests, the countervailing position, held by trade unionists and some members of the legal fraternity, was that consumers require protection, including from their own profligacy. The legislation as eventually passed maintained the appearance of an uneasy truce.

Conclusion

Initiatives such as Balboa and Avon represent but the most recent version of long-standing practices. From one point of view, they attempt to penetrate and capitalize on the financial dealings and activities of wage- and salary-earners, and grant recipients, in South Africa. Such initiatives are lauded by some for financial inclusion and democratizing credit, for banking the unbanked, and for opening up borrowing to formerly marginalized groups at the 'bottom of the pyramid' (Porteous with Hazelhurst 2004, p. 89). Others, suspicious of the greater inclusivity through which businesses and policymakers alike tap into informal economies or create linkages between the formal and informal, decry the 'regressive social and distributional effects' of incorporating informality (Meagher 2012, p. 10). Financial inclusivity, in short, can harm those included (Epstein 2005, p. 5). This latter approach would have it that a new and more insidiously 'marketized' version of capitalism is driving vulnerable people with few other livelihood options to help capitalists fleece their neighbours, and that they are thus complicit in the financialization of daily life. But how clear-sighted a reform package would be required to regulate and ameliorate something in which such complicit participation is already well established?

There were important distinctions between land and credit apartheid, and land and credit reform. Land arrangements were governed by far more specific, explicitly racially discriminatory laws, while credit arrangements have been governed by laws that seem, on the face of it, more global and less specifically South African. Reforming both proved difficult. If land reform was charged, at least on a symbolic level, with fixing all the problems of apartheid, then it is certainly the case that, even had these laws been accepted and implemented, it would have failed to do so. Time did not stand still: people had little wish to return to a forgotten past as pastoralists or cultivators. They wanted to move forward, they longed for – and many had already achieved – upward mobility alongside a modern, urbane and sophisticated lifestyle. Land reform nevertheless retains some force as a discourse: the state, having initially tried to curb the extent of restitution claims, has now seen fit to extend their validity into the ever-receding future. This has raised suspicions that the promised reforms are being used to defer other, more cogent demands for the restoration of citizenship.

Reforming credit is more complex, the behaviours requiring modification perhaps more puzzling. If *lenders*, for example, operate by the logic of the self-regulating free market, as they claim, then why is state regulation required to restrain them from offering products that will lead borrowers into penury, killing the goose that laid the golden egg? The assumption that *debtors* require protection (Wiggins 1997, p. 511), even from themselves, has resonated widely. For those whose aim was to become modern city dwellers rather than remaining in the country to benefit from land reform, but who had slender means for doing so, access to credit simultaneously enabled their aspirations while revealing their precarious character. The state was able to deliver democratic freedoms, but it could not – in a liberalized economy – deliver the means necessary to enjoy the life of the free person. The only way to get that was 'on tick'. Preventing such investment – not just in the frivolous consumerism over which much media hand-wringing has occurred, but also in valued long-term endeavours like higher education – would require not only that 'reckless lenders' stop extending loans but also that 'reckless borrowers' stop availing themselves of these.

Notes

1   This research was funded by a grant from the Economic and Social Research Council (award RES-062-23-1290), which I gratefully acknowledge. Opinions expressed are my own. The research, using participant observation and interviews, was part of a broader study on indebtedness and financialization. Thanks to those with whom I held discussions during my fieldwork. For comment on earlier versions of this paper I am grateful to participants in the Department of History Colloquia Series talk, University of British Columbia, and in particular to Dianne Newell who invited me.
2   SA Leads the world in living on credit, *The Times*, 3 June 2015.
3   Kevin Davie, Drowning in debt or rolling in riches, *Mail and Guardian*, 22 July 2011. Davie provides these figures but shows that they are subject to reinterpretation.
4   This and other quoted sections in the chapter are from a conversation with Xolela May, Knysna, 8 October 2008. A garnishee order is a court order 'instructing a garnishee (a bank) that funds held on behalf of a debtor (the judgement debtor) should not be released until directed by the court. The order may also instruct the bank to pay a given sum to the judgement creditor (the person to whom a debt is owed by the judgement debtor) from these funds.' http://www.finance-glossary.com/define/garnishee-order/612/0/g. The term bears no apparent relation to the more common meaning of 'garnish': to ornament or embellish.
5   'Low savings a concern says Minister Gordhan', *Engineering News*, 23 July 2009.
6   This is the technical legal term used for taking part of a salary before it is paid to the employee.
7   Frans Haupt, Pretoria, 3 September 2008.
8   A class action court case, brought to the Western Cape High Court, ruled against some of these practices in 2015, and the ruling was upheld by the Constitutional Court in 2016. The respondents found to have been using these techniques were drawn from among the new micro-lending companies. They included Mavava Trading, Onecor, Amplisol, Triple Advanced Investments, Bridge Debt, Las Manos Investments, Polkadots Properties, Money Box Investments, Maravedi Credit Solutions, Icom, Villa Des Roses, Triple Advanced Investments. The names give a sense of the fly-by-night character of those that had been gleaning rich pickings from the garnishee system. University of Stellenbosch Legal Aid Clinic and Others v Minister of Justice And Correctional Services and Others (16703/14) [2015] ZAWCHC 99 (8 July 2015), http://www.saflii.org/za/cases/ZAWCHC/2015/99.html#. For the 2016 Constitutional Court judgment, see http://www.saflii.org/za/cases/ZACC/2016/32.html.
9   For a similar practice in India, see Parry (2012).

10 Many traders were Gujarati-speaking Muslims from the Punjab or Jewish refugees from Russia and its borderlands (Cobley 1990, p. 43; Krige 2011, p. 137; Roth 2004, p. 62; Whelan 2011).
11 Gabriel Davel, personal communication. Gillian Jones, Revised financial regulation bill clarifies 'twin peaks' Business Day. 17 December 2014, http://www.bdlive.co.za/business/financial/2014/12/17/revised-financial-regulation-bill-clarifies-twin-peaks.
12 Similar legislation applies in the US, where 'Exemptions are created by statutes to avoid leaving a debtor with no means of support. For example, only a certain amount of work income may be garnished.' http://legal-dictionary.thefreedictionary.com/Garnishee+order.
13 Commentary on the National Credit Bill by Vincent Van Der Merwe, J.C. Grobler & Burger Inc., 5 August 2005, presented to the portfolio committee for Trade and Industry.
14 Existing legislation restricting the interest rate had been removed in 1992, ostensibly 'to open up the market for small borrowers': it 'removed price control on small loans' of under R6000 and with a duration of less than 36 months (Porteous with Hazelhurst 2004, pp. 77, 83). A cap was later reimposed at 44 per cent.
15 Balboa submission on the National Credit Bill, 28 July 2005, presented to the Portfolio Committee for Trade and Industry.
16 Avon Ladies try to make a living by direct selling of beauty products to friends and neighbours, and by recruiting them to the selling network in turn: Avon is 'the most successful direct selling empire in the industry' (Dolan and Scott 2009).

References

Adam, H., F. van Zyl Slabbert and K Moodley. 1998. *Comrades in Business: Post-Liberation Politics in South Africa*. Utrecht, The Netherlands: International Books.
Barchiesi, F. 2011. *Precarious Liberation: Workers, the State, and Contested Social Citizenship in Postapartheid South Africa*. Albany: State University of New York Press.
Beinart, W. 1979. European traders and the Mpondo paramountcy, 1878–1886. *The Journal of African History* 20 (4):471–86.
Beinart, W. 1986. Settler accumulation in East Griqualand from the demise of the Griqua to the Natives Land Act. In *Putting a Plough to the Ground: Accumulation and Dispossession in Rural South Africa, 1850–193,*. eds William Beinart, Peter Delius and Stanley Trapido. Johannesburg: Ravan Press.
Beinart, W., and P. Delius. 1986. Introduction. In *Putting a Plough to the Ground: Accumulation and Dispossession in Rural South Africa, 1850–1930*, eds William Beinart, Peter Delius and Stanley Trapido. Johannesburg: Ravan Press.

Bond, P. 2000. *Elite Transition: From Apartheid to Neoliberalism in South Africa*. London: Pluto Press.

Boraine, A., and M. Roestoff. 2002. Fresh start procedures for consumer debtors in South African bankruptcy law. *International Insolvency Review* 11 (1):1–11.

Crush, J. 1986. Swazi migrant workers and the Witwatersrand gold mines 1886–1920. *Journal of Historical Geography* 12 (1):27–40.

Dolan, C., and L. Scott. 2009. Lipstick evangelism: Avon Trading circles and gender empowerment in South Africa. *Gender and Development* 17 (2):203–17.

DTI (Department of Trade and Industry)/Reality Research Africa. 2002. Credit contract disclosure and associated factors. Pretoria.

DTI (Department of Trade and Industry). 2004. Consumer credit law reform: policy framework for consumer credit. Pretoria.

Ducastel, A., and W. Anseeuw. 2013. Agriculture as an asset class: Financialisation of the (South) African farming sector. Paper presented at conference on The Human Economy: Economy and Democracy, University of Pretoria.

Epstein, G.A. 2005. Introduction: Financialization and the world economy. In *Financialization and the World Economy*, ed. Gerald A. Epstein. Cheltenham: Edward Elgar.

Falk Moore, S. 2011. The legislative dismantling of a colonial and an apartheid state. *Annual Review of Law and Social Science* 7:1–15.

First, R. 1983. *The Mozambican Miner: Proletarian and Peasant*. New York: Macmillan.

Friedman, K.E., and J. Friedman. 2008. *Historical Transformations: The Anthropology of Global Systems*. Lanham, MD: AltaMira.

Graeber, D. 2011. *Debt: The First 5,000 Years*. New York: Melville House.

Hann, C., and K. Hart. 2011. Introduction: learning from Polanyi 1. In *Market and Society: The Great Transformation Today*, eds C. Hann and K. Hart. Cambridge: Academic Publishers.

.Hay, M. 2014. A tangled past: Land settlement, removals and restitution in Letaba District, 1900–2013. *Journal of Southern African Studies* 40 (4):745–60.

Hay, M. 2015. South Africa's land reform in historical perspective: Land settlement and agriculture in Mopani District, Limpopo, 19th century to 2015. PhD thesis, Witwatersrand University.

Haupt, F., and H. Coetzee. 2008. The emoluments attachment order and the employer. In *Employee Financial Wellness: A Corporate Social Responsibility*, ed. E. Crous. Pretoria: GTZ (Deutsche Gesellschaft für Technische Zusammenarbeit).

Haupt, Frans, H. Coetzee, D. de Villiers and Jeanne-Mari Fouché. 2008. The incidence of and the undesirable practices relating to garnishee orders in South Africa. Pretoria, GTZ (Deutsche Gesellschaft für Technische Zusammenarbeit).

James, D. 2007. *Gaining Ground? 'Rights' and 'Property' in South African Land Reform*, London: Routledge.

James, D. 2011. The return of the broker: consensus, hierarchy and choice in South African land reform. *JRAI* 17:318–38.

James, D. 2015. *Money From Nothing: Indebtedness and Aspiration in South Africa*. Palo Alto: Stanford University Press.

Jeeves, A. 1985. *Migrant Labour in South Africa's Mining Economy: The Struggle for the Gold Mines' Labour Supply, 1890–1920*. Kingston, ON: Queen's University Press.

Jeeves, A. 1990. Migrant labour in the industrial transformation of South Africa, 1920–1960. In *Studies in the Economic History of Southern Africa: Volume Two: South Africa, Lesotho and Swaziland*, eds Z.A. Konczacki, J.L. Parpart and T.M. Shaw. London: Routledge.

Kalb, D. 2012. Thinking about neoliberalism as if the crisis was actually happening. *Social Anthropology* 20 (3):318–30.

Kalb, D. 2013. Financialization and the capitalist moment: Marx versus Weber in the anthropology of global systems. *American Ethnologist* 40 (2):258–66.

Kaplan, M. 1986. *Jewish Roots in the South African Economy*. Cape Town: C. Struik.

Kear, M. 2013. Governing *Homo subprimicus*: Beyond financial citizenship, exclusion, and rights. *Antipode* 45 (4):926–46.

Krige, D. 2014. Letting money work for us: Self-organization and financialisation from below in an all-male savings club in Soweto. In *People, Money and Power in the Economic Crisis*, eds. Keith Hart and John Sharp. New York: Berghahn Books.

Langley, P. 2008. *The Everyday Life of Global Finance: Saving and Borrowing in Anglo-America*. Oxford: Oxford University Press.

Marais, H. 2001. *South Africa: Limits to Change – The Political Economy of Transition,*. 2nd edn. London: Zed Books.

Marais, H. 2011. *South Africa Pushed to the Limit: The Political Economy of Change*. London: Zed Books.

Martin, N. 2010. Class, patronage and coercion in the Pakistani Punjab and in Swat. In *Beyond Swat: History, Society and Economy along the Afghanistan-Pakistan Frontier*, eds Benjamin Hopkins and Magnus Marsden, 107–18. London: Hurst.

Martin, R. 2002. *The Financialization of Daily Life*. Philadelphia: Temple University Press.

Martino, E. 2017. Dash-peonage: The contradictions of debt bondage in the colonial plantations of Fernando Pó. *Africa* 87(1):53–78.

Meagher, K. 2012. The trouble that lurks beneath: globalization, African informal labour and the employment illusion. Paper delivered at ECAS conference, Lisbon.

Mosse, D. 2004. *Cultivating Development: An Ethnography of Aid Policy and Practice*. London: Pluto Press.

Murray, C. 1992. *Black Mountain: Land, Class and Power in the Eastern Orange Free State 1880s–1980s*. Johannesburg: Witwatersrand University Press.

Murray Li, T. 2010. Indigeneity, capitalism, and the management of dispossession, *Current Anthropology* 51 (3):385–414.

Palmer, R. 2001. Lawyers and land reform in South Africa: a review of the land, housing and development work of the Legal Resources Centre (LRC). Mimeo.

Parry, J. 2012. Suicide in a central Indian steel town. *Contributions to Indian Sociology* 46 (1&2):145–80.

Peebles, G. 2010. The anthropology of credit and debt. *Annual Review of Anthropology* 39:225–40.

Phillips, R.E. 1938. *The Bantu in the City: A Study of Cultural Adjustment on the Witwatersrand.* Alice: The Lovedale Press.

Porteous, D., with E. Hazelhurst. 2004. *Banking On Change: Democratizing Finance in South Africa, 1994–2004 and Beyond.* Cape Town: Double Storey Books.

Roth, J. 2004. Spoilt for choice: financial services in an African township. PhD dissertation, University of Cambridge.

Schapera, I. 1947. *Migrant Labour and Tribal Life.* London: Oxford University Press.

Schoombie, A. 2009. Access to formal financial services for South Africa's poor: Developments since 1990. *South African Journal of Economic History* 24 (2):131–56.

Schraten, J. 2014. The transformation of the South African credit market. *Transformation: Critical Perspectives on Southern Africa* 85:1–20.

Shah, A. 2010. *In the Shadows of the State: Indigenous Politics, Environmentalism, and Insurgency in Jharkhand, India.* Durham, NC: Duke University Press.

Stadler, A. 1979. Birds in the cornfield: squatter movements in Johannesburg 1944–1947. *Journal of Southern African Studies* 6 (1):93–123.

Trapido, S. 1978. Landlord and tenant in a colonial economy: The Transvaal 1880–1910. *Journal of Southern African Studies* 5(1):26–58.

van Onselen, C. 1996. *The Seed is Mine: The Life of Kas Maine, a South African Sharecropper, 1894–1985.* New York: Hill & Wang.

Walker, C. 2000. Relocating restitution. *Transformation* 44:1–16.

Wiggins, M.J.N. 1997. Rethinking the structure of insolvency law in South Africa. *New York Law School Journal of International and Comparative Law* 17:509–13.

Part III
Financialization

7

Infrastructure's Contradictions

How Private Finance is Reshaping Cities

Phillip O'Neill

Introduction

In 1776 Adam Smith nominated public works as one of only three responsibilities of government.[1] For the next two centuries the administrations of advanced capitalist societies and their citizens accepted this charter. The driver was the need for capitalism to access, on one hand, pools of labour of sufficient size and quality to ensure production of an increasing volume and diversity of goods and services; and, on the other, growing numbers of mass spending households with the aspiration for a good life. The modern city was the spatial instrument that enabled this access – with infrastructure providing the means for a city's economic and social life to be coordinated and synchronized successfully. By the mid twentieth century successful planning, financing and operation of urban infrastructure were testament to the evolutionary development and importance of the modern state. By then, infrastructure exceeded all other state portfolios, including defence, as the state's major field of endeavour.

Curiously, after such success, especially in underpinning capitalist expansion, this position within the state's capacity has now been dismantled and a reassembly of responsibility for urban infrastructure is underway across all nations of the world.[2] More curious, perhaps, it is a transformation that coincides with the greatest demand for infrastructure provision the world has ever seen (KPMG 2012a). This record demand has complex origins.

Money and Finance After the Crisis: Critical Thinking for Uncertain Times,
First Edition. Edited by Brett Christophers, Andrew Leyshon and Geoff Mann.
© 2017 John Wiley & Sons Ltd. Published 2017 by John Wiley & Sons Ltd.

First, in both advanced and developing nations, twentieth-century infrastructure is urgently in need of upgrade and maintenance (American Society of Civil Engineers 2015). What was built to underpin urban growth in the twentieth century – to supply energy, water and telecommunications, to move people and freight, and to guarantee natural resources flows in liveable habitats – requires expensive intervention to ensure a next century of operation. Second, technological change creates demand for investment in new infrastructure (HM Treasury 2013). For example, meeting demand for mobile telecommunications for both business and personal users requires the rollout of more powerful microwave and satellite networks. Likewise, advances in GPS and digital surveillance technologies enable new traffic management systems for better delivery of public transport services and more efficient allocation of road surfaces, for example by 'real time' adjustments to user rights and payments. Third, record rates of population growth and urbanization in emerging nations necessitate new infrastructure rollout at astonishing rates. Related, rapid growth in the production of consumer goods in the emerging economies means ongoing growth in the volume and distance of world trade links. Massive increases in container-ship capacity drive investment in ports, railways and motorways including in the old developed nations. Fourth, climate change and sustainability imperatives are generating new approaches to infrastructure across all aspects of energy, water, transport and telecommunications provision requiring investment in both installations (like solar and wind farms, and water recycling plant) and networks (like smart electricity grids and revamped water distribution systems) (KPMG 2012b).

Yet these demands come at a time when the fiscal power of governments is sorely diminished. The last quarter of the twentieth century was a tough period financially for governments across the world. Government revenues in advanced nations in particular had been stressed by long periods of economic stagnation since the 1970s. Then global financial crisis (GFC) in 2007–8 compounded the failure of governments to generate funds for public capital expenditure. Severe recession caused falls in taxation receipts while economic stimulus measures and bank bail-outs caused expenditure blow-outs. Public debt rose commensurately and government borrowing capacity fell (Reinhart and Rogoff 2010).

The combined incidence of these forces and events means the infrastructure problem now confronting governments is daunting. For the US, the American Society of Civil Engineers (2015) estimates the nation needs to spend US$3.6 trillion by 2020 for its infrastructure systems to survive in 'a state of good repair'.[3] Worldwide, the McKinsey Global Institute (2013) calculates that US$57 trillion of investments in transport (road, rail ports and airports), power, water and telecommunications infrastructure is required by 2030 simply to maintain current levels of GDP growth. Significantly, the amount – US$57 trillion – is more than the estimated total value of the world's existing infrastructure assets and 60 per cent more than was spent in these infrastructure categories in the previous eighteen years. Moreover, the McKinsey estimate does not include infrastructure expenditure which redresses the maintenance and growth needs of emerging economies, nor does it include the cost of infrastructure refit in order to meet pressing global environmental concerns, especially those relating to climate change.

A view of the economy that fails to understand its spatial composition would see this infrastructure problem as a fiscal problem. All that is needed according to this view is the restoration of structural balance to government finances so that revenue streams cover expenditures adequately (which in neoliberalist modes of governance would usually mean expenditure cuts rather than taxation increases) thereby generating savings for infrastructure investment and improving borrowing capacity. Yet, as world economies continue to struggle for growth and increases in taxation revenues dawdle, the infrastructure problem persists.

Does fiscal crisis mean an indefinite postponement of possibilities for the development of sustainable infrastructure systems in just, prosperous cities of the future, let alone the delivery of basic infrastructure to ensure the urban efficiencies that bolster economic growth, and jobs and income generation? Clearly there is a need for new approaches to infrastructure commissioning and financing. This chapter discusses the parameters for such developments and the ways these generate the tensions which are inherent in all contemporary urban infrastructure provision. One tension is that urban infrastructure is necessarily a collective endeavour such that the state's organizational and regulatory powers, including its property powers, are vital to infrastructure's existence; while at the same time, private institutions hold the capacity to finance infrastructure and, increasingly, to design, build and operate it. Another is that the citizenry of a metropolitan area sees infrastructure provision as essential to quality urban living, while private economic interests see infrastructure as necessary for assembling the means of production and distribution in the pursuit of private profit. As we shall see, these tensions are difficult to resolve and increasingly involve major dilemmas for public policy as the presence of the private sector expands even after major investor shakeouts during the GFC.

The section below, 'Infrastructure and the Crisis of Twentieth-Century Capitalism' revisits the way infrastructure provision and financing occurred in the post-war period, commencing with an explanation of successful infrastructure rollout in the immediate post-war years followed by some insights into the consequences of infrastructure privatization and sell-off as economic stagnation descended in the latter decades of the twentieth century. The next section, 'New Modes of Infrastructure Financing' builds on this historical analysis by charting the dimensions for the emergence of private sector payers in infrastructure investment and delivery, and the emergence of an infrastructure investment class. 'The Consequences for Cities' section then evaluates these changes within the context of the role of infrastructure in the actual constitution of the city itself. The discussion here is an attempt to reposition infrastructure away from it being a fiscal event – however much this is a crucial issue – to open our understanding of infrastructure as a key material entity in the construction of daily urban life and thereby a major – if not *the* major – determinant of the quality and fairness of urban living. Finally, 'Conclusion: A New Politics of Infrastructure' makes some conclusions as to how this knowledge might be mobilized for better infrastructure futures.

Infrastructure and the Crisis of Twentieth-Century Capitalism

Infrastructure was not magically invented by interventionist twentieth-century governments deploying Keynesian fiscal strategies. The need for a comprehensive infrastructure base to human settlement was well advanced in the eighteenth and nineteenth centuries with roads, railways, bridges and canals delivered by both state and private means. Indeed, infrastructure provision was a precondition to the advance of industrial capitalism. Almost all infrastructure assets in cities are thoroughfares or passages or transportation corridors. They were essential to early capitalism for the assembly of labour and other factors of production and for the distribution of products to markets. Infrastructure enabled larger numbers of people to live in towns and cities. Supply chains could reach further into the countryside, and the growth of markets enabled the specializations and divisions of labour that fuelled the growth of profits and the accumulation of capital. Infrastructure made commercial sense.

Infrastructure investments deepened and broadened in the decades following the Great Depression and the Second World War. Keynesian fiscal strategies enabled the major commitment of government capital spending to infrastructure which was supported enthusiastically by private enterprise because of the commercial benefits that flowed, and by a wider electorate keen to see urban and suburban amenity enhanced as western cities grew in size and area. In addition, new forms of passageways became available for investment including electricity grids, urban transit systems, telecommunications and broadcasting systems and air routes. As a bonus, infrastructure investments carried very large economic multiplier benefits.

Importantly, infrastructure investment in the expansive, post-war period also created or expanded state capacities. This included the development of urban planning as a government portfolio, a ministry which became essential to the rollout of post-war suburbanization with infrastructure spending delivering enormous value gain to land subdivisions and related property developments. Also important was the creation of a new state entity, the utility, an ownership and organizational vehicle that became the natural, commonsense way that infrastructure was delivered.[4] Here civic society, such as through its representation on the boards of utilities, gained a direct say in infrastructure design and rollout. This presence helped affirm the legitimacy of the utilities' claim on monopoly market power and their independence from formal politics, while entrenching guaranteed flows of often fully-funded finance (Beasley 1988).[5] Ironically, the invention and operation of the utility by the state made the corporatization and privatization of infrastructure a much easier task decades later. It's hardly surprising that so many of the world's largest corporations today are former electricity and tele-communications utilities.

By the late 1960s, the western world had clarity about the importance of public investment in the passageways and conduits that enabled efficient flows of people, goods, water, energy and information in cities. An instinct for infrastructure had developed. Public works became sensible. The idea of infrastructure as the substrata of a city which integrated its parts and enabled its efficient functioning had been conceived and mobilized. Moreover, the financial and engineering capacities for infrastructure's initiation and perpetuation were well established. Two things were important here:

- Infrastructure became a *subsumed political process*. Infrastructure was funded off the public sector balance sheet without political con-testation. Then its procurement and provisioning were handled by state utilities with a corps of planners and engineers who, by and large, operated autonomously in evaluating urban infrastructure needs and finding technical and financial solutions. Importantly within this apparatus of infrastructural venture the risk of failure was accepted within the wider ambit of the state.
- Infrastructure also became a *subsumed socio-spatial process*. As we explore in the section 'The Consequences for Cities', each new infra-structure item was bundled, synchronized and sequenced with pre-existing items. Often this occurred where there was major new

construction or reconstruction of parts of a city such that bundling, synchronizing and sequencing objectives could be planned for; but it also occurred in circumstances where change was incremental such that obliging such objectives happened as a matter of course.

Then, together, these subsumed processes were generative of two important (socio-spatial) political aspirations. One was the norm of universal access, meaning not so much that the use of infrastructure became free and unfettered, but that infrastructure acted in ways that enhanced – rather than diminished – the interaction of individuals and other entities across the domains and functions of a city. The second was that the complementarities and externalities generated by an infrastructure asset, and those generated by the operation of the asset in concert with other infrastructure assets, existed as services and benefits that were available for use at less than full cost, such as the operation of a train to bring late night concert goers home to the suburbs; or as non-commodified externalities such as the use of parklands created in nature reserves set aside as part of urban flood mitigation schemes.[6]

Widespread goodwill surrounding urban infrastructure provision in the post-war period in western nations meant that the fiscal consequences of infrastructure spending could be kept transparent and did not incur political risk. The public saw infrastructure spending as a necessary and desirable part of the progress of a city. A positive approach to infrastructure spending was rewarded by income and employment multiplier effects especially from construction, productivity improvements across the economy, and from widespread access to complementarities and externalities. In turn, enhanced economic growth rates yielded restorative public revenues. A virtuous cycle of provision and reward was generated.

From around the 1980s, however, consensus around the purpose and provision of infrastructure in western nations unravelled. The forces of change were many and complex, coalescing around issues of declining sovereignty in the face of globalization, slowing economic growth rates, constraints on states' fiscal capacities, and a loss of confidence in the effectiveness of the state apparatus in the supply of essential public services. A common response was to source resuscitation programmes from the handbooks of neoliberalism based around the deployment of private capital in market-controlled systems of production and distribution. Agreement about infrastructure's commissioning, design, funding, ownership and operation, and about its role in planning and building cities, and in aiding commerce, became confused at first and then vigorously contested. The shift to private procurement, financing and operation took about three decades in most advanced nations, such was the deeply entrenched nature of the utilities and state-owned enterprises in the urban infrastructure sector. In these decades there have been failed experiments in the commissioning and operation of private infrastructure assets alongside numerous instances when the transfer of public assets to private hands occurred without the state maximizing commercial value or devising appropriate regulatory structures to minimize practices such as price gouging in circumstances where little or no supply choice existed. There were also bold experiments in heavily geared investment structures and synthetic financial products, many of which collapsed during the GFC, like similar experiments in other investment classes. Yet despite market failure, and the recklessness and dishonesty that exaggerated the intensity of the GFC in the infrastructure sector, there is now an established presence in urban infrastructure of private financing, procurement, construction, operation, perhaps even regulation. A discussion of the nature of this presence follows in the next section, 'New Modes of Infrastructure Financing'.

New Modes of Infrastructure Financing

Beyond the general argument about the appropriateness of an expanded role for the private sector in a modern economy, there are particular considerations about the role of private ownership and private investment practices in urban infrastructure. On the one hand, the development and operation of infrastructure creates an integrated urban platform for collective use by a city's businesses and householders. This function generates infrastructure's public good characteristics. On the other hand, infrastructure is changing into a set of discrete private assets each with attributable revenue streams and self-interested management arrangements that seek, ultimately, to maximize returns on private investment (O'Neill 2009). This function creates a private investment domain for infrastructure. Significantly, governance arrangements for infrastructure have become burdened with the competing expectations that arise from these different functions. Rightly, there is a continuing project within a city's polity and its technocracy that sees infrastructure as needing to be planned and operated such that the functions and parts of a city are integrated, efficient and sustainable. At the same time, however, interest among private asset managers in a continuing, wider urban management project is steered towards those matters directly bearing on the profitability of the asset for which they are responsible. We return to this issue later.

As we have seen, infrastructure has a number of characteristics which have enabled its transformation from a set of assets designed to enable

Table 7.1   Infrastructure assets as an investment class

Characteristic	Effect on investment quality
Revenues are generated under natural monopoly conditions or under regulated conditions where competition is minimized or eliminated.	Competition risk is minimized.
They are relatively demand inelastic.	Returns are recession proof compared to competing asset classes such as public equity stocks and commercial property.
They are relatively stable spatially and materially and subject to only modest levels of technological change.	They are relatively long-term investments with little need for the development of liquidity (tradability) qualities.
Their operation is readily observable and can be metricized without resort to opaque algorithms and forecasting.	Their investment qualities are relatively transparent which minimize transactions costs.

the flows of things in and around cities to devices capable of extracting private earnings from these flows. That said, it should be understood that infrastructure as an investment class is now much more than the textbook depiction of infrastructure as a public good or natural monopoly.[7] Infrastructure as an investment class includes more market-sensitive (and potentially competitive) assets like ports, airports and telecommunications installations; organizational entities that own infrastructure operating rights like an infrastructure services corporation; packages of urban services and amenities – quasi-infrastructure assets – that are demand-inelastic like parking meters and funeral service providers; and debt contracts and securities with rights over revenue streams from infrastructure user tolls. As investment items within their own asset class, then, infrastructure assets are now characterized by a new set of qualities (see Table 7.1), ones related as much to elements like liquidity, risk and yield than to contributions to urban efficiency and amenity; and where exclusive rights can be assigned to the urban thoroughfares that enable the flows of people, materials, energy and information; and to revenues collected from users of these flows.

So, the transformation of infrastructure assets into operations that generate cash returns on a regular, predictable basis has produced a new, and now stable, asset class. Post-GFC and stripped of much of its exotic material, the asset class has the additional merit of being backed by

Table 7.2   Major non-state players in the infrastructure sector

Group	Examples
Privatized former state-controlled utilities and enterprises	Deutsche Telekom AG, Électricité de France S.A., Terna S.p.A, Japan Railways Group, Royal KPN N.V., Auckland Airport, PSA International Pte Ltd, (formerly Port of Singapore Authority), Thames Water Utilities Ltd
Capital aggregators especially the global banks and their agencies	J.P. Morgan, Macquarie Bank, Morgan Stanley, Deutsche, Goldman Sachs, UBS, Mitsubishi UFJ, Société Générale, Banco Santander
Savings aggregators especially pension funds, insurance and sovereign wealth funds	OMERS, OTPP, CPP, Australian Super, CalPERS, ABP, CAF, QIC, APG, Hanwha, Future Fund, Abu Dhabi Investment Authority
Multidisciplinary corporations	Brookfield Asset Management, Inc., Ferrovial, S.A., Abertis Infraestructuras S.A., Bouygues S.A., Meridiam SAS. Lend Lease Ltd
Specialist infrastructure funds	Infracapital, Global Infrastructure Partners, Highstar Capital, KIAMCO, IFM

genuine and observable value providing the type of instrument that was once present in an intermediated financial world, but became increasingly absent from a financial scene disrupted by the disintermediation of securitized and collateralized debt relationships. Infrastructure products, then, offer an alternative to synthetic financial products seen as lacking recognizable substance. Moreover, infrastructure is an investment product with fairly simple metrics – traffic counts, usage rates, daily toll earnings – capable of being understood and verified through desktop checks by an average funds manager. Being seen as capable of generating stable, long-term yields has made infrastructure products a natural fit for financing from superannuation and pension funds rather than, say, from marketable debt instruments.

The rise of the infrastructure investment sector has also generated a new set of institutional players within a financialized capitalism, and empirical evidence shows they now dominate the ownership registers of the major assets for each of infrastructure's subsectors. Table 7.2 lists these major players.[8]

First, there are those companies spawned by the privatization of major utilities in the 1980s and 1990s, chiefly from the electricity, telecommunications sectors in the UK and Western Europe. Many of these companies have subsequently grown through mergers and acquisitions into truly global corporations.

The next group are the major banks, especially the large banks from North America, Europe and East Asia. These banks take multiple positions in infrastructure investing. In their role as capital aggregators they contribute both debt and equity capital to major infrastructure deals. They are also key advisers to both investors and governments in transacting new and brownfields infrastructure investments, and they also operate independent investment funds.

A major competitor to the banks are the savings aggregators, including the major pension funds, insurance companies and trusts, and some of the larger sovereign wealth funds. The savings aggregators are attracted to major infrastructure investment because of the close match between the long-term stable returns from infrastructure investments and the long-term liabilities to savings holders. A key recent feature of investment practices of this group is a desire for direct investment, meaning control over the management of the assets, in order to avoid the substantial fees that are payable to other agencies when management is outsourced, say to funds or specialist corporations (Clark and Monk 2013).

An intriguing group of players in infrastructure investors are the multidisciplinary corporations, typically large, listed, former construction companies enticed by the gains available from taking equity positions in large infrastructure projects and direct roles in construction contracts and project finance initiation. Such multidisciplinary presence in the infrastructure sector enables these tightly organized global companies to maintain strong relationships with other key players in the sector including financiers, governments, planners and operators.

Finally, there are the specialist funds, in many ways a legacy group from the first wave of private infrastructure investing, but with major market presence still, due to their control over key assets secured in early wave privatizations.

Figure 7.1 then shows how this group of players has manoeuvred itself into positions with advantaged opportunities for ownership of infrastructure assets and control over streams of infrastructure revenues. The figure sketches the passage of organizational structures in infrastructure provision in advanced nations starting from the circumstances in the nineteenth century as the nation state evolved in structure and importance with growing capacity to fund, construct and operate major urban infrastructure assets. Experimentation in organizational structures during the Great Depression and World War II years, and immediately thereafter, saw the creation of state-owned utilities which housed the full range of disciplines – fiscal, engineering, regulatory, operational and so on – needed for running integrated urban infrastructure systems. Interestingly, it was the late twentieth century modernization of the public sector that saw the utilities converted into state-owned

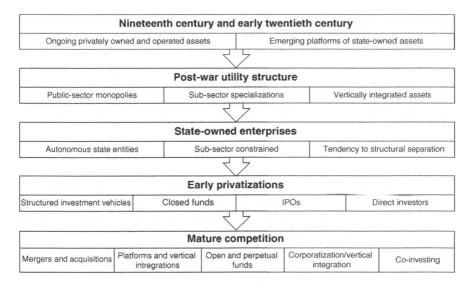

Nineteenth century and early twentieth century	
Ongoing privately owned and operated assets	Emerging platforms of state-owned assets

Post-war utility structure		
Public-sector monopolies	Sub-sector specializations	Vertically integrated assets

State-owned enterprises		
Autonomous state entities	Sub-sector constrained	Tendency to structural separation

Early privatizations			
Structured investment vehicles	Closed funds	IPOs	Direct investors

Mature competition				
Mergers and acquisitions	Platforms and vertical intregrations	Open and perpetual funds	Corporatization/vertical integration	Co-investing

Figure 7.1   Evolution of organizational structure in the infrastructure investment sector.

enterprises as states sought to enhance efficiency and effectiveness within their bureaucracies and operations. Not only did the utilities embrace corporate-like structures, the organizational and cultural change involved meant that as fiscal and ideological pressures grew for the privatization of publicly owned assets the utilities, now in enterprise mode, had become ready-made for sale into private hands.

The figure then tracks the modes of private sector entry into infrastructure operation during the early periods of privatization, in most nations stretching from the 1980s to the end of the century. The main forms were structured investment vehicles, or SIVs, which enabled tight contractual ties between infrastructure revenues and investors; closed funds, being small pools of private investors committed to a long-term ownership position given the illiquid nature of most infrastructure assets; IPOs, or public offerings, designed to shift a well-structured state enterprise into a listed company with a broad ownership base, often to placate political opposition to the privatization; and, by direct investment such as acquisition by a private pension or insurance company, sovereign wealth fund or foreign entity.

Finally, the figure shows how a more stable investment profile has emerged. While there remain many players in infrastructure investment worldwide, the presence of large numbers of players is confined to subsectors where investment lots are small – such as municipal water schemes, small regional transport facilities, and PPPs (public private partnerships, also known as P3s) involving public buildings without substantial value. In respect to large scale, expensive infrastructure assets, be they greenfield or brownfields, then, the investment field is now highly concentrated among a few large global players. These players are typically from one of four groupings: corporations, direct investors, large funds and large financial institutions (sometimes called capital aggregators). Figure 7.1 shows that their strategies typically involve:

- mergers and acquisitions, as larger players seek to capture the best assets;
- platforms and vertical integrations, whereby investors seek to build complementarities across their asset holdings by concentrating in specific subsectors (platforming) or by rebuilding anti-competitive vertical structures such as, in the case of electricity, activities around generation, distribution and retailing;
- open and perpetual funds formation, designed to maintain ongoing control of large infrastructure assets given the availability of such choice assets is finite in number;
- corporatization and vertical integration, particularly involving infrastructure constructors and infrastructure operators keen to establish stronger investor positions in the sector; and
- co-investing, whereby large investors with access to surplus debt or equity funding are keen to take additional investment positions alongside other investors who may well be competitors in other circumstances.

The purpose of understanding the infrastructure investment sector as a maturing investment class is to expose how large financial institutions and aggregators, and global corporations of various forms, have become central controllers of not just the great twentieth century state-owned instrumentalities but also the repositories of the expertise and financial power that need to be mobilized whenever sizeable new projects warrant commissioning. So privatization has not simply involved a loss of control by governments of previously owned and run infrastructure assets, privatization handed to the private sector and its associates the organizational capacity, including for design and financial responsibility, to build and operate new assets.

The Consequences for Cities

Economic, environmental and social outcomes flow from the privatization and financialization of the infrastructure sector. Central to these outcomes is the nature and extent of the ways the internal flows and functions of cities are driven by the need to abide the financial contracts underpinning urban infrastructure; and how the broader economic performance of a city relies increasingly on the networked advantages that privately engineered inter-urban infrastructural flows create.

The concerns here are threefold. One is for the material efficiency of the city. How can a city's mix of private and public infrastructure assets be brought together to enable a city to function effectively? Consider the coordination task of Transport for London, the agency responsible for public transport in the United Kingdom capital. TFL manages 24 million daily tube journeys, plus another 300,000 journeys on the Docklands Light Rail and countless others on sixty overground trains. Then it supervises the daily operation of 9,000 buses on 675 routes. Then it coordinates coach travellers such as through the Victoria Coach Station which serves 14 million customers per annum involving 240,000 coach departures to more than 1,200 destinations across the UK and 400 in Europe. These public transport infrastructures must then be coordinated, especially when they are roads-based, with flows of car traffic in London amounting to over 10 million journeys per day, cycle journeys numbering now over 500,000 journeys per day, and pedestrian movements of over 6 million journeys per day.[9] Consider then that these people flows intersect materially, as well as in social and temporal fashions, with freight flows, and with water, energy and telecommunications flows, and that these need also to be coordinated in layout and operation. Coordination and management of the entire infrastructure platform, in any major city, then becomes a complex and difficult but highly necessary task. So aligning these public interests with stakeholder interest in the maximization of private investment returns adds a substantial new layer of considerations to an already monumental logistical challenge.

A second concern is the capacity of a city to meet a set of collective aspirations each of which is heavily dependent on the city's infrastructure platform. These are the desires for a city to generate:

- widespread prosperity typically through an economy that can generate appropriate levels of employment and incomes, and reserves (savings and wealth) for the future;
- social and cultural enrichment from daily neighbourhood interactions and city-wide major events; and
- biophysical environments that are diverse and replenishing.

The third concern is highlighted by Graham and Marvin (2001) who highlight the role of a city's infrastructure platform in generating social justice. Where private interests dominate, they argue, there is unfair betterment via improvements in the social and technological mobilities of the more affluent and the more powerful, while households which are poorly located or poorly remunerated suffer deepening disadvantage. Where infrastructure provision is partial – Graham and Marvin say 'splintered' – and commodified then aspirations for universal access are curtailed, and publicly-accessible positive externalities are diminished, such that the poor are not just denied access to the basic circulations of the city, but also from the life of the city itself.

So what can we say about the consequences of infrastructure privatization for the operation of cities? In the section 'Infrastructure and the Crisis of Twentieth-Century Capitalism' above, we saw how post-war (public) infrastructure provision became a subsumed political process alongside a subsumed socio-spatial urban development process. These are valuable attributes that need revisiting. The ultimate challenge, then, for a privatized, financialized infrastructure provision model is to ensure the continued functioning of the spatial platform that is provided by what we might call *infrastructure-in-aggregate*. The remainder of this section develops this idea.

An effective, fairly-functioning infrastructure platform of a city can be assessed by four criteria: universality, bundling, access and positive externality. First, to the principle of universality. This asserts the aspiration for infrastructure – an idealistic one, certainly – to be available in all circumstances. The principle of universality can be explained by thinking about an urban residential street. The simple idea that dwellings should have common set-backs, neatly along a street, is rarely contested for the reason that living along a street ensures that the residents in every household receive an entitlement to basic, life-enabling services because of where their dwelling is positioned. The street was developed as more than a thoroughfare for vehicles and pedestrians. It is the most common expression of the civilized notion of a public right to water, sewerage, power and telecommunications. These systems are installed in the corridor created by the common set-back, running along every street, past every dwelling. In a parallel sense, the street also gathers households into formations that demand, by their very presence, the expectation that every household is entitled to use of the services provided by a city's infrastructural platform. So a clear political statement is built physically into every urban street.

Of course, where daily life is not configured by the street, such as in rural areas or among remote indigenous communities, then the principle of universality has sometimes to be argued for. Yet while the response of government and the state apparatus has never been fully adequate, the legitimacy of the demand for basic services based on a citizen having an entitlement to them is usually acknowledged, at least in advanced nations. Universality, then, has had historical acceptance, even if only as work-in-progress.

Beyond the spatial services arising from street-side urban living, the extent to which universality can permeate other urban infrastructure provisioning has generally been determined by the adoption (or not) of the principle of bundling. Bundling refers to the delivery of items of infrastructure in braided channels so that solving the problem of access to one infrastructure item usually guarantees access to others, meaning infrastructure items can operate in complementary fashion. Like universality, bundling comes naturally when infrastructure is provided at the scale of the street. The point of having a common delivery device – the street – is that basic infrastructure items can be rolled out in a common pattern such that every household has coordinated use of the whole package. Bundling also operates at wider scales. The best examples are city-wide public transport and arterial road systems which perform similar networking functions to each other. An urban rail or arterial road network in post-war suburbanizing cities linked daily street life to productive, commercial and social life in other parts of the city. Important here was the way suburban rail networks also bundled workers into wider urban divisions of labour. Each morning, suburban railway stations gathered local workers for train journeys to CBDs, commercial centres and industrial zones. Later, cars and buses replaced trains as the preferred transport means, and arterial roads became the key infrastructure for bundling urban life; but that's another story.

Together, universality and bundling then produced and enforced the principle of access. Universality and bundling normalized a citizen's right to access basic services, and to access places of employment from new suburbs; with these suburbs designed explicitly to provide affordable housing options involving large scale public as well as private housing development schemes and tenure options.

The way urban life became integrated by the coincidence of universality, bundling and access is intriguing. Together, universality, bundling and access produced a rhythm in daily life for urban communities.[10] A normal day was bundled time-wise into sequences that were over-determined by the nature and role of the urban infrastructure with which householders intersected. The timing of the train or bus was linked precisely with the timing of the working day, the morning peak hour inscribing the start of the working day and the afternoon peak determining its end. The school day was then nested inside the paid-work schedule. So too the opening hours of shops were part of daily and weekly rhythms around infrastructure use and operation. Bundled, well-used infrastructure produced parcels of coordinated and sequenced time with urban activities and household routines in concordance, at least overtly.

Importantly, the accord between the state and the citizenry about the funding and rollout of infrastructure came not just from infrastructure items being characterized by universality, bundling and access but from the free positive externalities that were generated by the infrastructure package. For each and every water and sewerage customer, for example, there is, concomitantly, a safe, hygienic city. From the power stations and electricity grids constructed to light and heat homes, there is also vast access to industrial power by manufacturers. From a road system that steers traffic onto efficient arterial roads there are (or there should be) quiet, safe suburban streets. And for (and, indeed, because of) every public transport user, there is cleaner air and there are fewer greenhouse gas emissions. Of course, the wider community also endures negative externalities from infrastructure. Power poles might be unsightly, and unsafe for wayward drivers. Roads and railways generate noise and smells. Sewerage systems require expensive treatment plants and, often, environmentally-concerning discharge. Yet negative externalities are usually assuaged by positive ones and by the overwhelming benefits that come from universality, bundling and access. As part of the bundled urban infrastructure package, then, negative externalities are tolerated to the extent that they are often rendered invisible. Critically, positive externalities, the benefits that flowed across the city and across generations, reinforced the instinct that made infrastructure part and parcel of building the post-war city.

The question for a privatized, financialized infrastructure, then, is whether individual infrastructure assets, now deliberately made discrete and unbundled, are capable of delivering to a city an infrastructure platform (infrastructure-in-aggregate) where universality, the building of urban flows, the ideal of access, and the generation of net positive externalities emerge as a direct consequence. The conclusion section which follows looks at the possibilities.

Conclusion: A New Politics of Infrastructure

The threat to urban life from infrastructure privatization comes, then, when prioritizing actions to generate long-term financial returns from infrastructure investment puts at risk infrastructure's capacity to generate the raft of services and their externalities once seen as intrinsic to public sector capital investment. The direction beckoned by privatized and financialized infrastructure now seems likely to be dominated by assets with these characteristics:

- they are owned and managed privately;
- they are organized into discrete functional and organizational entities;

- they have fully monetized costs and returns;
- they have known and assigned financial and operational risks; and
- they have bespoke regulatory arrangements designed primarily for the financial viability of the asset rather than the functioning efficiency of a city.

Political opposition to such a future, however, should not involve nostalgic pursuit of a post-war infrastructure formation, for we should remember, among other things, that this formation was rolled out to accommodate high growth, high energy consumption, suburbanizing cities. The world's cities now have different needs so different models of infrastructure provision are needed. That said, there are elements of infrastructure design and organization that merit resurrection as principles to guide infrastructure platforms for twenty-first-century cities that are just, prosperous and sustainable. The principles of universality, bundling, accessibility and being generative of positive externalities are four worthy of re-adoption.

Yet adopting these principles in today's diverse, continually changing cities requires new approaches. Infrastructure assets need to be designed so that they are flexible in both function and capacity through time. A transmission network designed for high current, high voltage electricity generated from large coal-burning installations should probably be designed for reduced load into the future, and therefore for a loss in investment value. Similarly, water systems might be designed to accommodate growing volumes of non-commercial water storage and flows, so their owners should deliberately anticipate value loss. Less centralized, more distributed energy and water systems might also become more dynamic if larger numbers of innovative supply and use modes were able to be incorporated into them. And perhaps it makes sense for these innovations to come from for-profit ventures using non-state sourced finance in order to attract best quality technologies and organizations. And perhaps new privately devised systems of coordination designed explicitly to build synchronizations, sequences, access and complementarities across an infrastructure platform might be better delivered by a for-profit provider. It would be heartening to be able to point to a capability within the state to deliver the innovation and management efficiencies needed to meet these aspirations. But we know the state apparatus is a battered thing, meaning we need to look to hybrid public–private solutions in order for the state to recover contemporary relevance (see Freudenberg 2015).

The point is that the delivery of public good by the urban infrastructure platform is essential, but it is not necessary that this be done by the public sector. Of course, only the public sector can provide the regulatory power to ensure infrastructure provision in cities; and the public sector is the organizational place to enact the priority of wider benefits over financial returns both for infrastructure assets and for infrastructure-in-aggregate. The state has the primary role to play as regulator and organizer. In such a context, dynamic efficiencies from private, for-profit ventures can have a genuine role in driving successful twenty-first-century cities.

Acknowledgements

The fourth section 'The Consequences for Cities' in this chapter contains material previously published in O'Neill (2010). Support for the research was provided by two grants from the Australian Research Council, DP1096324 and DP130104319, for which the author is very grateful. The author would like to thank the Bartlett School of Planning, University College London, for hosting his visit there in 2015, where this chapter was written, and the editors for their helpful comments and encouragement.

Notes

1 According to Smith, 'The third and last duty of the sovereign or common-wealth, is that of erecting and maintaining those public institutions and those public works, which though they may be in the highest degree advantageous to a great society, are, however, of such a nature, that the profit could never repay the expense to any individual, or small number of individuals; and which it, therefore, cannot be expected that any individual, or small number of individuals, should erect or maintain. The performance of this duty requires, too, very different degrees of expense in the different periods of society' (Smith 1977, book V, chap. 1, part 3).

2 Torrance (2008; 2009) maps progress in the privatization and financialization of infrastructure pre-GFC. Her work sets out the investment landscape from which this chapter attempts to build.

3 http://www.infrastructurereportcard.org/ accessed 6 May 2015.

4 For a history of the New South Wales experience see Beauchamp, 2006.

5 There is an important distinction between capital works financing and its funding. Financing refers to access to finance (cash or credit) for expenditure on a project. Funding is the source of payment – the cost of borrowing – for access to finance. Funding for public infrastructure is usually from taxation revenue or from direct user fees or tolls.

6 For a discussion of the efficiency and productivity gains that flow from infra-structure, and the positive externalities, including improved economic and social equity, that can be generated by infrastructure provision, see Fourie (2006).

7   See O'Neill 2014 (which draws heavily on Lakoff 1987) for discussion of the primary characteristics that determine the admission of things like a bridge, a railway line and a water and sewage system into the infrastructure category. The discussion therein contrasts with the narrow discussion in academic economics, (say by Coase 1974 in his famous lighthouse article), where infrastructure is portrayed as a type of economic good involving natural monopoly provision.

8   Sources for this discussion include material from the online infrastructure investment data base at www.preqin.com.

9   Information from www.tfl.gov.uk, accessed 5 June 2015.

10   Here I borrow from Amin and Thrift (2002) who point to the ways that urban life produces vital transivity and rhythm. Transivity and rhythm in my view are fashioned by urban infrastructure, although Amin and Thrift attribute transivity and rhythm to softer technologies and practices. But the words I prefer, because they are more deliberate, more invoking of engineering solutions than I think Thrift and Amin had in mind, are synchronizations and sequences.

References

American Society of Civil Engineers. 2015. *Report Card for America's Infrastructure 2013*. American Society of Civil Engineers: Reston VA.

Amin, A. and N. Thrift. 2002. *Cities: Reimagining the Urban*. Oxford: Polity Press.

Beasley, M. 1988. *The Sweat of their Brows*. Sydney: Sydney Water Board.

Beauchamp, C. 2006. *Parliament, Politics and Public Works: A History of the New South Wales Public Works Committee 1888–1930*. Sydney: New South Wales Parliamentary Library.

Clark, G.L., and A.H.B. Monk. 2013. The scope of financial institutions: Insourcing, outsourcing and off-shoring. *Journal of Economic Geography* 13:279–98.

Coase, R.H. 1974. The lighthouse in economics. *Journal of Law and Economics* 17:357–76.

Fourie, J. 2006. Economic infrastructure: definitions, theory, empirics. *South African Journal of Economics* 74:530–55.

Freudenberg, G. 2015. Contemporary relevance, comrade: Gough Whitlam in the 21st century. *The Whitlam Institute's Commemorative Gough Whitlam Oration 2015*. Sydney: Whitlam Institute.

Graham, S., and S. Marvin. 2001. *Splintering Urbanism*. London: Routledge.

HM Treasury. 2013. *Planning for Economic Infrastructure*. London: HM Treasury.

KPMG. 2012a. *The Great Global Infrastructure Opportunity*. Amsterdam: KPMG International Cooperative.

KPMG. 2012b. *Cities Infrastructure: A Report on Sustainability*. Amsterdam: KPMG International Cooperative.

Lakoff, G. 1987. *Women, Fire and Dangerous Things: What Categories Reveal About the Mind*. Chicago: University of Chicago Press.

McKinsey Global Institute. 2013. *Infrastructure Productivity: How to save $1 Trillion a Year*. London: McKinsey Global Institute.

O'Neill. P.M. 2009. Infrastructure investment and the management of risk. In *Managing Financial Risks: From Global to Local*, eds G.L. Clark, A.D. Dixon and A.H.B. Monk, 163–88. Oxford: Oxford University Press.

O'Neill. P. M. 2010. Infrastructure financing and operation in the contemporary city. *Geographical Research* 48:3–12.

O'Neill, P.M. 2014. How infrastructure became a structured investment vehicle. In *Engaging Geographies: Landscapes Life Courses, and Mobilities*, eds M. Roche, J. Mansvelt, R. Prince and A.Gallagher, 29–44. Newcasstle: Cambridge Scholars Press.

Reinhart, C.M., and K.S. Rogoff. 2010. Growth in a time of debt. *American Economic Review* 100:573–8.

Smith. A. 1977 [1776]. *An Inquiry into the Nature and Causes of the Wealth of Nations*. Chicago: University of Chicago Press.

Torrance, M.I. 2008. Forging glocal governance? Urban infrastructures as networked financial products. *International Journal of Urban and Regional Research* 32:1–21.

Torrance. M. 2009. The rise of a global infrastructure market through relational investing. *Economic Geography* 85:75–97.

8

The Financialization
of Nature Conservation?

Jessica Dempsey

'Biodiversity Crunch'

In the midst of the 2008 subprime crisis, I was in London to attend a
Biodiversity and Ecosystem Finance conference on the edge of the financial district. In the comfort of my hotel, I watched BBC commentators
debate how the UK could avoid a major recession, with the government
suggesting a cut (eventually enacted) to the value-added tax to stimulate
consumer spending. It was an especially anxiety-ridden time throughout
North America and Europe; amidst all the upheaval I selfishly worried
that the financial pale would ruin my fieldwork: would anybody attend
a conference like this? One paradox of our time is that, while recessions
can be good for reducing environmental impacts like greenhouse gas
emissions, they also tend to reduce the signal of environmental issues on
the political radar. In the end, around fifty people came to the conference.
Most were European professionals – environmentalists/conservationists,
heads of corporate social responsibility (CSR) at firms/banks, environmental financiers and government bureaucrats.

Although everyone had gathered to discuss how to extend private
capital into the conservation of species and ecosystems, no one could
avoid talking about the broader financial crisis. The result was a strangely
recombinant discourse. The keynote speaker, British conservationist
David Bellamy, coined the terms 'biodiversity crunch' and 'Amazonian
crunch', drawing together ideas of the credit crunch of the time with
ongoing biodiversity loss.[1] Later that day, an executive from Rabobank,
a large Dutch bank deeply involved in international agricultural lending,

Money and Finance After the Crisis: Critical Thinking for Uncertain Times,
First Edition. Edited by Brett Christophers, Andrew Leyshon and Geoff Mann.
© 2017 John Wiley & Sons Ltd. Published 2017 by John Wiley & Sons Ltd.

quipped that 'bankers are an endangered species', producing many chuckles and knowing glances around the room.

The conference aimed to address a second crisis: the problem of a world that is becoming more the same, a planet becoming less lively, less colourful, less diverse. Since the beginning of the twentieth century, about 75 per cent of the genetic diversity of agricultural crops has been lost (Food and Agriculture Organization 2010); in the past forty years, the abundance of moths and butterflies on the planet has declined by as much as 45 per cent (Dirzo et al. 2014). MacKinnon (2010) describes a 'ten per cent world'; the planet now has only 10 per cent of the biological variety and abundance it had before the mass culls and extractions that have marked the period from early imperial capitalism to the present. The scale of losses is so vast some now refer to it as the sixth extinction, a term that draws attention to the five previous mass extinction episodes in the Earth's historical record, events profound in pace and scale of extinction. These events decimate both fit and unfit species, effectively restructuring the biosphere (Kolbert 2009). Some scientists say we are living through the sixth extinction now, a mass extinction event considered to be caused by human-induced changes (Ceballos et al. 2015).

In spite of this context of crises both financial and ecological, participants' discourse was confident, even celebratory. Ten years ago, said Michael Kelly, chair of the London conference and head of CSR for the professional services company KPMG-UK, such a conference would have been 'impossible'. In Kelly's view, the Biodiversity and Ecosystem Finance event itself was a major achievement. The subprime crisis of the moment, constituted in disastrous innovations of finance (particularly mortgage-backed securities), should not lead to widespread questioning of the role that private, interest-bearing capital should have in solving the problems of the sixth extinction. Rather, the financial crisis was positioned as a moment to pass through. And then, when the cash started flowing again, the question to be grappled with would be how to make more of that cash flow in the direction of mitigating ecological problems. Kelly's comments and the conference title – Biodiversity and Ecosystem Finance (two earlier conferences with the same name were held in New York in 2008 and 2009) – demonstrate some sort of growing relationship between biodiversity, ecosystems and finance. But what is the nature of this relationship? Is it changing? If so, how? And where might it be going?

The Rabobank speaker delivered one version of an answer. He began with an image of a large sugarcane plantation in Brazil. The plantation caused deforestation, he explained, and *his bank funded it*. This is the most obvious relation between finance and ecosystem change: financial firms (public and private) provide equity or debt finance for corporate agricultural and extractive activity that leads to land use changes. In this relation, capital produces natures that are more alike: that is, monocultures. This process recently has been intensified and accelerated by trade and finance liberalization, but from the early imperialist and colonial flows of capital/trade, finance has been an undeniable part of the story of the sixth extinction. When land and water are treated primarily as vehicles for delivering maximum return (or as dumping grounds), the dangers for humans and non-humans are many.

My interest in this chapter, however, is in a more specific question: Is the mitigation of ecological impoverishment – in particular the conservation of biological diversity – becoming 'financialized'? Scare quotes are required, because the answer depends on how one understands financialization. The term is used in multiple ways: to describe, inter alia, relationships between finance and what are previously considered 'non-financial' institutions or processes; the contemporary nature of a global capitalism that increasingly relies upon financial profit; or changes to people, firms or organizations who increasingly understand themselves and relate to each other as financialized subjects (for an overview see Christophers [2015]). The term seems to be used at times as a verb, describing a process, but also as an adjective, to describe the character of contemporary capitalism, as in 'financialized capitalism'. In this chapter, I retain a broad definition, exploring both processes and outcomes. If we theorize the innovations and expansion of finance as crucial in rescuing investors from a low-growth, low-profit economy (Arrighi 1994; Blackburn 2006; Harvey 2005), the particular question I am interested in is whether finance is seeking to profit from efforts to ameliorate the sixth extinction. If so, how? Are these innovations linked to crises of over-accumulation? Again, how so? Finally, are financializing processes so transformative that we might claim to be in the midst of a widespread financialization of conservation? And what might be at stake in such a turn?

Sullivan (2013) has argued that the 'domain of finance capital' is coming to 'control environmental conservation' (p. 207) because 'business and finance sectors, in collaboration with conservation organisations, conservation biologists and environmental economists, are engaging in an intensified financialisation of discourses and endeavors associated with environmental conservation and sustainability' (p. 199). Financialization also appears as a key attribute of what several scholars call neoliberal conservation. For example, Bram Buscher and Robert Fletcher (2015) link together regimes of global accumulation with what they call regimes of conservation, linking the contemporary, twenty-first century financialization accumulation phase with what they call a 'fictitious conservation' phase that involves the marketization and financialization of conservation.

These authors argue that we are living through a broad regime shift, entering an era of 'accumulation by conservation'.[2] They draw from Sullivan (2013), noting this phase involves 'interest in a variety of innovative financial instruments including carbon markets, species and wetlands banking, environmental derivatives and suchlike' (p. 286), and cite as examples the United Nations Framework Convention on Climate Change's Kyoto Protocol and the development of the global carbon market. For Buscher and Fletcher, these international arrangements are signals of an epochal change towards a distinctively green regime of accumulation wherein capitalism conserves the conditions needed for its own long-term reproduction. They claim 'conservation is increasingly central to accumulation processes in contemporary capitalism' (p. 277), part not only of an attempt to save capitalism from itself, but also of 'a last-ditch effort to recover profit when concrete commodity markets have exhausted their potential' (291). Buscher and Fletcher thus link the growing interest in the financialization of conservation to broader crises in capitalist accumulation – where capital, facing reduced profits from production, seeks out greater yield from 'speculative ventures of all sorts' (O'Connor 1994, cited in Buscher and Fletcher 2015, p. 291).

Concerns about financialization also preoccupy left think tanks and critical social movements. For example, Friends of the Earth International (FoE) (2014) opposes the 'financialization of nature', which the organization perceives as rapidly proceeding in the form of cap and trade proposals, biodiversity banks and natural capital bonds. An FoE policy paper connects global economic and environmental crises: 'Today, as a way out of the financial crisis, companies and markets are rapidly directing their greed towards Nature and all its resources, because it represents a new market necessary to overcome previous crises caused by the same system' (p. 9). Malaysia-based Third World Network has produced a similar policy paper, written by Sullivan (2012), while a wide-ranging group of organizations, including FoE, Ecologistas en Acción and Carbon Trade Watch, recently produced a short film entitled 'Stop the Takeover of Nature by Financial Markets'.[3] It lays out the stakes of 'submitting nature to the logic of capital':

> With new property rights a new phase of enclosures and extraction of natural resources and biodiversity grabbing is taking place. So people are kicked off their lands, communities and nature are destroyed. And the Global South becomes the loser. All those engaged in the Green Economy and financial markets become the winners. This green economy is no solution to the crises ... It essentially offers no solution to problems we face.

In short, many scholarly and activist circles use the term financialization as shorthand for ominous and dark times, a signal of wider, deepening capitalist logic in all spheres of life.

I too am concerned by what could happen if more of life's essential building blocks are traded, accumulated and controlled by the private sector. At the same time, there is debate over the extent and impacts of this financialization. Political ecologists Brockington and Duffy (2010, p. 480) argue that conservation has 'hardly been involved in the production of value through financialization'. Brockington and Duffy note the use of debt for nature swaps and some 'interchanging of personnel' between financial institutions and conservation organizations (e.g., the president and CEO of The Nature Conservancy is a former director of Goldman Sachs).[4] However, to Brockington and Duffy, these 'are peripheral developments' (p. 480). More recently, scholars note the lagging or non-existent market in so-called market-based conservation (Fletcher and Breitling 2012; Pirad 2012) and the marginality and small scale of accumulation in conservation overall (Dempsey and Suarez 2016).

So are we in fact living in a moment of widespread financialization of nature, an epochal regime shift to 'accumulation by conservation'? In this chapter, after exploring some key definitions, I lay out five distinct but connected research questions that I hope can help the critical academic community systematically assess the extent and effects of the financialization of biodiversity conservation. The possible influences of finance can take many forms, and those of us concerned with these trends need to be more specific about the forms financialization may or may not be taking. As this suggests, the chapter by no means definitively answers the question that began this paragraph. After reviewing some of the existing evidence, I end by discussing the lack of financial capital circulating in the project of biodiversity conservation and the implications of this.

From Economization to Marketization to Financialization?

The generation of more resources for conservation – namely resources to flow from North to South – is the main sticking point of every single negotiation of the Convention on Biological Diversity. Large conservation organizations also call for more funding. Recent work by the World Wildlife Fund (WWF) and Credit Suisse found that the amount of additional funds needed to conserve biodiversity was US$300 billion per annum (Credit Suisse et al. 2014).

Conservation finance tends to refer to 'raising, borrowing, investing and managing' capital to support conservation (Clark 2007, p. xiv). This capital includes public finance; public finance has always been the largest source of funding for conservation.[5] When people talk about the financialization of conservation, however, they are generally not referring to a growth in public finance or charitable giving. Rather, the term tends to refer to an expansion of private, return-seeking (especially interest-bearing) capital in economic, social and environmental relations. As international conservationists repeat (over and over again), the funding gap is highly unlikely to be filled with public finance; reports emphasize a crucial need for private, return-seeking finance (Bishop et al. 2008; Credit Suisse et al. 2014; Huwyler et al. 2014).

As discussions of financialization have proliferated in academic literatures, the term's meaning, as Brett Christophers (2015) argues, has stretched. For Christophers, financialization has 'by now largely lost any coherence that it previously enjoyed: increasingly standing only for a vague notion of "the (increased) contemporary importance of finance"' (2015, p. 184). He goes as far to suggest that 'its enrolment today risks raising more questions than it answers' (ibid). In both academic and activist writings on neoliberal natures, financialization can seem to be a catch-all phrase, referring to any attempt to turn environmental mitigation efforts into a money-making opportunity.

One way to define financialization more specifically is to compare it with 'economization' and 'marketization', two terms that refer to different (although not separate) processes. Economization generally refers to the extension of economic logic, practices and calculation into new areas (Çalışkan and Callon 2009). It does not mean market making, but it does involve assigning economic value to previously unpriced or unvalued entities or processes in order to establish some way of measuring possible choices, as in cost–benefit analysis. It includes the valuation of ecosystem services and the creation of economic incentives, including many payments for ecosystem services (PES) schemes.[6] Marketization refers to the creation of markets – opportunities for buying and selling commodities – where they did not exist before. Examples include wetland banking and of course the market in carbon, especially forest or ecosystem carbon. Marketization is, in effect, what Kathleen McAfee (1999) meant when she coined her widely cited phrase 'selling nature to save it'.

At its most general, financialization refers to the growing influence of financial actors, interests, and institutions – an influence that is marked in relation to other parts of the economy. A hallmark use of the term financialization often chronicles how 'non-financial' businesses – those that produce goods and services – are leaning more on financial channels to earn profit (Krippner 2005), participating in debt markets and in pricing, distributing and hedging risk (Blackburn 2006). Some, such as Martin (2002), theorize that the policy and governance shifts of recent decades have produced individuals as financialized subjects who must take responsibility for wisely investing their own assets (including stocks of human, social and financial capital) or else face an insecure or precarious future.

But how much influence do financial actors, interests and institutions have on biodiversity conservation, historically and at present? I now turn to five research questions – sub-areas of inquiry that might help researchers and activists gauge this level of influence more specifically. Some literature exists on each area; however, as this five-part overview demonstrates, further research is needed.

Investigating the Financialization of Conservation

Are biodiversity conservation relationships financializing?

A basic study of financialization could begin by characterizing the changing terms of the relationships between financial firms/institutions and conservation. This could involve analysing the composition of boards and staff structures (financiers are far more likely to appear on the boards/staff of conservation organizations than conservation representatives are to appear within financial firms) and identifying any joint projects or initiatives. Are these financial firms being courted by conservationists or vice versa? Are these firms being offered green PR for their work, or do the organizations demand changes to financial firm operation as a requirement for the relationship? Addressing these questions involves identifying interconnections and assessing how deep they go.

While more systematic mapping of these relationships is needed, even a basic scan of the landscape reveals that individuals with financial backgrounds are clearly participating in the governing of biodiversity conservation organizations. As noted earlier, a former managing director of Goldman Sachs became the president and CEO of The Nature Conservancy (TNC) in 2008.[7] The head of the first major international initiative to value nature, The Economics of Ecosystems and Biodiversity Initiative (TEEB), was a former banker – Pavan Sukhdev, who has a long history in finance, with fourteen years at Deutsche Bank.

What is the significance of these individual relationships joining financiers and conservation? For one, they suggest a *continuation* of international conservation as an elite project.[8] Is there evidence that financiers are entering the conservation world in rates greater than in the past? And what might these relationships mean for the way that conservation is conducted? A systematic study showing these relationships over time is needed to more precisely assess these connections and their influence.

At the institutional level, it also appears that financial institutions are collaborating more actively with conservation institutions; the Biodiversity and Ecosystem Finance events with which I opened the chapter signal such a change. Michael Kelly, head of CSR at London-KPMG and chair of the London event, suggested that these conferences reflect a shift, noting that only ten years ago it would have been impossible to get these people in the same room. While this particular series of conferences did not continue after 2009, others sprung up: the Natural Capital Forum in 2014 and 2015, the Global Landscape Forum in the summer 2015, and US-based Conservation Finance events in 2014–2016. Participants in these events included representatives from conservation organizations, corporations, international organizations and financial institutions. The Natural Capital Forum, for example, is partially funded by Alliance Trust, a major UK investment bank, while Credit Suisse supported the Global Landscape Forum and the Conservation Finance conferences.

In addition to events such as these, there are also signs of material partnerships. At the forefront of such connections continues to be the Nature Conservancy, which in 2014 created its impact investing unit, NatureVest.[9] With founding sponsor JP Morgan, the mission of NatureVest is to 'create and transact investable deals that deliver conservation results and financial returns for investors' (NatureVest 2016). The group aims to create investment products aligned with TNC's goals and raise capital for them, as well to grow the area of conservation investing more broadly.[10] Meanwhile, Credit Suisse is also emerging as a participant in conservation finance. In 2014 it partnered with WWF and McKinsey to produce the report *Conservation Finance: Moving Beyond Donor Funding Towards an Investor-Driven Approach*. Credit Suisse also helped organize an expert workshop (hosted by the Federal Reserve Bank of San Francisco in January 2014) with the world's leading conservation finance specialists and held a Conservation Finance event in its own New York boardrooms in 2015. Other NGOs have partnerships with financial firms. For example, Flora Fauna International has worked with Merrill Lynch, Australian bank Macquarie and the International Finance Corporation. And, as is to be expected, financial firms donate to conservation organizations: for example, TD Bank supports The Nature Conservancy; the Bank of America supports Conservation International. If you apply to open a WWF Bank of Americard Cash Rewards Visa credit card account, WWF receives $100 from Bank of America.

What this array of connections suggests is that alongside growing partnerships between conservation organizations and multinational corporations (MacDonald 2010) are growing relationships between conservation organizations and financial institutions. But there is still a need for systematic research into these relationships to understand what has changed, remains the same, and to understand what is at stake. What kind of influence could these deepening relationships effect? Why are financial firms participating: to green their brands, or perhaps to explore new investment strategies?

Is conservation discourse financializing?

At the 2009 Biodiversity and Ecosystem Finance conference in New York, participants identified the lack of common, convincing language as a key barrier to the mobilization of private sector capital for ecosystem conservation. For example, one investment banker commented on the use of the term 'bank' in mitigation and conservation banking. She said such terms are extremely confusing for those in the investment community: 'A lot of funds, they have analysts that cycle through, one day looking at energy, another day on paper, then on retail. They go on cycles. The term "bank" throws them for a loop. It loses time, causes confusion.' Yet, as *Guardian* columnist George Monbiot opined in 2014, more financially understandable ways of speaking about biodiversity have grown over the last decade (although they have much longer histories):

> Sorry, did I say nature? We don't call it that any more. It is now called natural capital. Ecological processes are called ecosystem services because, of course, they exist only to serve us. Hills, forests, rivers: these are terribly out-dated terms. They are now called green infrastructure. Biodiversity and habitats? Not at all *à la mode* my dear. We now call them asset classes in an ecosystems market. I am not making any of this up. These are the names we now give to the natural world.

There is no doubt that the language of ecosystem services and natural capital are growing in use among a wide variety of actors.

The twenty-nine financial firms that signed the Natural Capital Declaration (NCD) know what natural capital and ecosystem services mean (or at least their corporate social responsibility departments do). The NCD is a document committing signatory insurers, banks and investment managers to contribute to making 'reforms needed to create a financial system that reports on and ultimately accounts for the use, maintenance, and restoration of natural capital in the global economy'. It defines natural capital as comprising the 'Earth's natural assets (soil, air, water, flora and fauna)' and the 'ecosystem services resulting from them, which make human life possible'. These services are 'worth trillions of US dollars per year in equivalent terms and constitute food, fibre, water, health, energy, climate security and other essential services for everyone' (United Nations Environment Programme Financial Initiative [UNEP FI] and Global Canopy Programme 2012). Terms like ecosystem services and natural capital are proliferating in use in governments and within conservation organizations – for example, use of ecosystem services in the decisions of the CBD grew sharply over the 2000s (see Dempsey 2015). Yet there is more work to be done to *systematically* map the discursive shift, especially from a critical perspective, to better understand its driving forces. Does this shift stem from a growing influence of finance? Or it is driven by scientists? And if the latter, why are they embracing such economistic rhetoric?

And what is at stake? On the surface, the language seems to help firms and governments understand what ecosystems do and what negative effects might result from ecosystem degradation. Probing deeper, however, critical scholars worry about this shift. Sullivan (2010) describes the rise of ecosystem services as another instance of an 'imperial ecology,' given that ecosystem services – as metaphors, scientific practices, models and policy approaches – are largely formulated by experts located in institutions of the global North and aim to serve 'transcendental corporate capital and finance' (p. 119). She asks, 'What knowledge and experiences are being othered and displaced through the parlance and practice of ecosystem services?' (p. 23). Norgaard (2010) considers the framework of ecosystem services reductive, narrowly defining ecological problems as ones of 'stock and flow', 'blinding us to the ecological, economic, and political complexities of the challenges we actually face' (p. 1219). Monbiot (2014) argues that this discursive turn is 'effectively pushing the natural world even further into the system that is eating it alive'. Sullivan (2013) makes a similar point: '[t]he construction of nature as a "service provider" is a significant conceptual move enabling financial investment in measures of, and markets for, nature conservation' (p. 205). Social movement and activists also voice similar concerns (e.g., International Indigenous Peoples Movement for Self Determination and Liberation 2013; World People's Conference on Climate Change and the Rights of Mother Earth 2010).[11]

Yet calling ecosystem services like carbon sequestration an 'asset class' does not make an investable asset class by financial firms, managers and bankers (as I discuss in the following section). This also opens up other research questions about the sources and drivers of this turn: What is driving it? While concepts like ecosystem services and natural capital seemingly make it easier for capital and the state to see and understand and apprehend nature's work, it is important to note that (1) these are conceptual and discursive turns animated by often heterodox economists and ecologies, not only driven by accumulative or financial firms/interests/logics (Dempsey 2015; Gomez-Baggethun et al. 2010), and (2) their meanings and applications are not homogenous, stable or uncontested

(Barnaud and Antona 2014; Dempsey and Robertson 2012; Kull et al. 2015). In the critical literature there is a tendency to present these concepts as though they were created by firms looking for new sites of accumulation as a salve for falling rates of profit; however, the history of these concepts and their slow uptake, especially in regards to their accumulative potential, suggests that the 'why' question needs more work, as well as more specificity in the effects.

Is biodiversity conservation a site of financial investment and yield?

Now to the juicy question, the one I think most people mean when they talk about financialization: To what extent is interest-bearing capital involved in mitigating the sixth extinction, in investing equity and debt into conservation? Answering this question requires study of the size, scale, geographies and overall character of actually existing capital circulating in nature conservation and directed towards the mitigation of the biodiversity loss. We need to study more than discourses of or ideas for new biodiversity investment vehicles; we need to chart and understand flows of real capital in such instruments, at the macro level (the sector as a whole) and also within specific funds and projects, including those that fail.

Failure is a big part of the story to study and tell – the marketization of nature conservation is not easy. Challenges are rife. The focus most recently is on making revenue streams that can fund land and marine protection by selling the ecosystem services that flow from biodiversity, especially forest carbon sequestration but also water purification. This is an area I am actively researching. So far, our analysis shows that attempts to make commodities out of standing forests for carbon sequestration are not performing all that well; the business of selling nature to save it is not booming (Dempsey and Suarez 2016). For example, the market in forest carbon transacts only $250 million a year – the size of one Wal-Mart. For well over a decade, conservationists have been waiting for a robust carbon market to emerge, one that allows for avoided-deforestation credit production and sale. Proposals for forest-backed bonds have also been floating around for that same decade. But 'securitizing nature to save it' is difficult without an underlying asset and revenue flow to securitize (perhaps akin to an attempt to securitize mortgages when there is no demand for houses). It is difficult to make money off conservation and thus, just as difficult to make a robust asset class for investors – one with an attractive risk-return profile and good inflation protection, as well as historical data to base investment decisions on.

One example clearly illustrates this failure. In 2008, Canopy Capital, a private equity firm aiming to invest in what it calls the 'eco-utilities' of tropical forests, purchased the rights to calculate and market the eco-system services of a patch of Guyana's tropical forests, the 900,000 hectare Iwokrama Reserve. Funded by ten high-net-worth backers, this self-titled 'pioneering deal' speculated that a financial return would follow eventually from the sale of 'eco-utilities' provided by the forest. The speculation failed; no buyers for the eco-utilities emerged, not even for the carbon sequestration. The project also faced difficulties with local people and the government when it sought to assign legal rights to the ecosystem services. Canopy attempted to put in place a forest-backed bond to further finance its conservation work, but that initiative also was unsuccessful. As such, investors lost their shirts. There are success stories, especially in the voluntary carbon market. But the small amount of private finance moving in these circles tends to be more willing to take lower return or less liquidity (in other words, not market rate) – capital, very often, from high-net-worth investors less concerned about taking a haircut with their investment.

However, there are new models emerging. Take, for example, Althelia Ecosphere, the darling of the conservation finance world. Althelia is a €105 million closed-end fund launched in 2011 and due to mature in 2021. It invests in agroforestry and sustainable land use and claims that its returns are market rate. Investors are mostly quasi-public institutions like the European Investment Bank, the Dutch development bank FMO, FinnFund in Finland and the Church of Sweden, as well as the David and Lucile Packard Foundation. In addition to these investors, in 2015 the Fund and Credit Suisse issued Nature Conservation Notes, debt instruments that generated €15 million of finance from non-institutional investors.[12] Althelia's latest investment is a €7 million commitment toward protecting 570,000 hectares of natural forest in Peru. The project site includes national park reserves, and the investment will restore a 4,000 hectare degraded buffer zone around these parks. The plan is to eventually produce deforestation free cocoa that will create jobs for local farmers and generate four million tonnes of certified carbon emission reductions. While financial returns are meant to be market rate, Althelia is backed by a USAID guarantee which halves the financial risk associated with the projects, a classic example of the collectivization of private sector risk.

Furthermore, Althelia – the much-celebrated model of new conservation finance – does not have a revenue stream that is entirely in forest carbon credits. Rather, the fund aims to generate a mix of revenue that prominently features sustainable commodities. This is a trend also noted in a study conducted by NatureVest (2014) (the partnership between The

Nature Conservancy and JP Morgan), which found that 66 per cent of private finance invested in something called conservation ($1.9 billion over five years) is actually directed to what they call 'sustainable food and fibre projects' (like green commodities). These new models emerging in conservation finance suggest that private capital holders lack faith in the environmental credit market, especially where it interfaces with land use. These financial funds, like Althelia but also Moringa and SLM Australia Livestock Fund, focus more on creating traditional commodities (such as food and forest products) 'sustainably', with carbon or other ecosystem service credits as a cherry on the top (Althelia 2016; Moringa Fund 2016; Partners in Sustainable Land Management 2016).

Even these innovations in conservation finance remain limited in scale. They are relatively small boutique operations in the finance world: for example, the Althelia Climate Fund closed at €105 million, Moringa Fund at €51.4 million, the SLM Livestock fund at $75 million AUS. The question asked in this world is how to make the projects going on 'investable, scalable, and repeatable', as the introduction to a recent Conservation Finance (2016) conference in New York City explained. There is once again active talk of forest or landscape bonds to generate finance for funds and projects like Althelia, but so far it remains only talk. In short, while more research is needed to track the scope and scale of existing and emerging developments, early work suggests that the opportunities for interest-bearing capital in the asset class 'biodiversity conservation' remain limited.

Are risks from biodiversity loss and ecosystem change becoming financialized?

Loss of biodiversity may pose significant risks to firms, such as the impact of pollinator declines on agricultural firms, or perhaps regulatory or reputational risks that might come from businesses that impact biodiversity. If firms are not pouring capital into funds involved with biodiversity conservation, might they instead be willing to invest in insuring themselves against the risks associated with biodiversity loss? This question has two elements. First, are risks of biodiversity loss or ecological change being hedged and/or adapted to? Answering this question means identifying finance practices and/or products that allow firms to manage their risk from changing ecologies, akin to ones focused on climate risk mitigation such as insurance (Johnson 2013; Knudson forthcoming). Such financialization is not focused on mitigating or fixing environmental problems but rather on extending (and in some cases profiting from) insurance to firms or people to protect them from these changes.[13] Research into this biodiversity-related risk mitigation could involve studying new insurance products and interviewing investors to understand if and how firms are recognizing and managing these risks. I am not aware of any products focused on insuring firms or individuals from risks of biodiversity loss. There are, however, relationships forming around this issue: for example, reinsurance giant SwissRe is collaborating with The Nature Conservancy to understand whether incorporating ecosystems into insurance industry models will improve the assessment of disaster risk assessment and provide finer grain assessment of the costs and benefits of 'nature based adaptation', such as restoration of mangrove forests on coastlines (The Nature Conservancy 2016).

These activities and partnerships relate to the second element of my question regarding biodiversity risks: Are financial firms becoming aware of the risks they face from biodiversity loss (not simply adapting to them, as per above)? And if so, what are they doing to mitigate these risks? Over the past decade, think tanks, international organizations and NGOs have created principles, frameworks, tools and instruments to make financial institutions aware of the risks of biodiversity loss for project investments with the hope of spurring investment in conservation. These groups have developed a growing number of technologies and tools meant to help financial firms assess their exposure to operational and reputational risks stemming from changing ecologies (see Dempsey [2013] for an overview). If, for example, a financial firm is an Equator Principles signatory, its lending (bridge and project), project finance, and advisory services are subject to the International Finance Corporation's (IFC) performance standard six, which aims to protect and conserve biodiversity by setting out a minimum standard for the responsible management of environmental and social risks in project finance (International Finance Corporation 2012).[14] Any project financed by IFC or other Equator Principle institutions is required to avoid damaging critical habitats – areas with high biodiversity value. Such a standard requires that these financial firms have some way of knowing if a project or client is working in or near a critical habitat. One tool, the Integrated Biodiversity Assessment Tool, aggregates data about critical habitats. For a fee, corporate or financial firms can anonymously access this data to see if a project will have impacts on a critical habitat site, which ideally might lead to rejection for project finance, or more likely, to attempts to reroute or mitigate (the IFC performance standard does allow for offsets). Founding sponsors of the tool include the Bank of America, BP, Cargill, Chevron and JP Morgan.

Such tools are speedily proliferating. In 2008, the Business Social Responsibility Network published a report comparing seven different tools that assess multiple ecosystem services (Waage et al. 2008) and, in

2014, BSR produced a report titled *Making the Invisible Visible*, which listed over 120 tools (Waage and Kester 2014). The aforementioned Natural Capital Declaration has developed a 'Soft commodity forest-risk assessment' tool that allows financial firms to understand the risks to which they may be exposed if they provide financing for companies whose activities contribute to deforestation.

How are we to understand the incorporation of biodiversity loss into risk assessment calculations? They are most certainly attempts to produce a less friction-filled, risky world for corporate transactions, as attempts to secure the conditions for capitalist reproduction. As with the issues identified throughout this chapter, more research is needed to understand what these principles, assessments and tools do in the world: Are they changing the way business and finance do business? If so, to what extent? Bracking's (2012) research provides a helpful model; she studied the environmental, social and governance impact assessment systems of fifteen development finance institutions (DFIs). She concludes that these assessments have a 'high degree of pseudo-scientific complexity' and 'limited influence on firm behaviour' or investment decision-making, most effectively working to confirm 'the legitimacy, authority and power of financiers' (2012, p. 281). Furthermore, there is much capital circulating that is not held to these 'pseudo-scientific' assessments; for example, most private equity funds have 'no systems of environmental or social safeguarding and do not monitor the impact of the companies they invest in' (Bracking 2012, p. 274). Bracking goes on to suggest that the growth of private equity approaches to investment, especially natural resource extraction, can be in part explained by the way that they are less easily held to account by local communities and campaigners. Private equity funds 'do business from afar, with investors pooling their funds, and then Fund Managers invest in other companies, who may then invest in yet others, or set up a corporate entity with the special purpose of actually building, mining or drilling' (p. 275).

Evaluating how firm and investor behaviour changes as they join principles and utilize tools is challenging. As one UNEP Financial Initiative employee told me in an interview, most companies do not want to release the data on how they actually implement policies on biodiversity loss; 'It is difficult', he said, 'to assess how they implement this in practice.' Yet a 2015 World Economic Forum report states quite clearly that we are still a long way from widespread financialization of environmental risks, noting that while environmental risks are increasingly recognized and acknowledged, there is 'little progress', and 'governments and businesses remain woefully underprepared' (World Economic Forum [WEF] p. 21).

Are conservation organizations and actors becoming financialized subjects?

Existing critical literatures examine financial influences on personal subjectivities: Martin (2002) suggests that everyday life is becoming financialized; other research aims to understand the extent to which individuals, or perhaps organizations understand themselves and act as 'entrepreneurial investor subjects' (Hall 2012, p. 405). A similar line of inquiry now arises in studies of the financialization of biodiversity: Are conservation organizations and actors changing their goals and identities as they become entangled in relationships with financial sector? Are 'profit-oriented conservation subjectivities' (McGregor et al. 2014) emerging? Are they replacing the more moralistic orientations that initially gave rise to conservation?

Some critical scholars worry about the effects of these finance–conservation entanglements. Sullivan (2014, p. 29) suggests that the discourse of natural capital is 'creat[ing] patterned orders of thought and truth in the world', 'binding nature with economic concepts and structures' while at the same time severing nature from social-ecological relationships and other conceptions of value. Fletcher and Breitling (2012) suggest the advance of a kind of 'neoliberal environmentality' (p. 408), wherein conservation policy 'demonstrates an overarching emphasis on motivating behavioral change through incentives, monetary or otherwise' (p. 405), as opposed 'to formal regulations, ethical injunctions, or assertions concerning the essential nature of things' (p. 405). These critical insights still require further study, particularly to parse out the links between discursive changes, new partnerships, the advance of new conservation subjects, logics and mentalities and their actually existing effects on conservation goals and outcomes. This kind of work would add specificity and empirical weight to these critical perspectives.

One approach might be to look closely at changing patterns of funding by multilateral institutions, for example, that of the Global Environment Facility (GEF), a financial mechanism created out of the 1992 Rio Earth Summit with the mandate to finance global environmental benefits in the Global South. Funded by the North, the latest installment of the GEF (known as GEF 6) will deliver US$4.43 billion to projects in the Global South that create global environmental benefits (e.g., biodiversity conservation, climate change mitigation) between 2014 and 2018. GEF 6 represents the first time that the GEF is allowed to *loan* funds to private sector initiatives in its inaugural 'non grant' stream. While funds in this programme total just US$110 million (over the same four-year period), the initiative signals that GEF is indeed conducting itself more like a return-oriented financial institution, and perhaps even more so is encouraging the development of for profit conservation initiatives. Are other multilateral donors or foundations asking conservation actors to create 'investable' projects too?

And if so, what is the effect on conservation? Something like a financialized conservation subject is certainly desired by the sounds of the reports produced by financial firms dabbling in this space which explicitly state that there is a lack of entrepreneurial or business culture within conservation organizations (Conservation Finance Alliance 2014; Credit Suisse et al. 2014). The reports chastize these organizations for remaining too focused on handouts and donations. The Conservation Finance Alliance (2014, p. 6), for example, observes that 'the majority of activities in the environmental sector are not focused on financial sustainability, let alone generating profits or returns'. But is this a problem conservation organizations are actively trying to solve? Are there more business school graduates working in conservation organizations, for example, more people coming from financial organizations? Are new staff or board members promoting or facilitating significant changes in the organization, like in organizational mandates or resource priorities? And, of course, all of this has a geography: Where in the world are financialized conservation subjects appearing? Why?

And, returning to the sixth extinction, is the very goal or subject of conservation shifting as a result of growing financial relationships? Elsewhere I've written about how a growing emphasis on economic logic may be changing the very subject of conservation away from biodiversity towards a more targeted approach to some species and ecosystems (Dempsey 2016). While presenting a methodological challenge, it seems crucial to explore if and how some non-human lives are becoming more or less investable (or expendable) as financial firms become more involved in conservation. For example, the word biodiversity is absent from a 2016 report released by Credit Suisse with the title *Conservation Finance from Niche to Mainstream: The Building of an Institutional Asset Class*, a report released in collaboration with the International Union for the Conservation of Nature – the world's largest and most prominent biodiversity conservation organization. But what might this elimination of biodiversity mean for how conservation is materializing on the ground?

The Financialization of Nature Conservation?

Overall, my research over the past few years has been animated by my own vacillation between one pole, where I sense that the world of nature conservation is rapidly financializing, transforming into an arm of financial institutions such as JP Morgan, and another pole, where I see that very little is happening, that conservation continues along as in the past, largely marginal to global capital flows (except where people and communities successfully organize to oppose developments making investors very worried). I waver between a sense that 'control of environmental conservation' is shifting 'into the domain of financial capital' as Sullivan writes (2013, p. 207) and an increasing sense that financial capital is largely indifferent towards environmental conservation. Firms like JP Morgan and Credit Suisse are now carrying the torch of conservation finance, publishing reports and even creating investment products. But within these firms, I suspect that conservation finance hardly registers. On their websites, information about these projects is buried within the corporate responsibility departments, not in the main operations of the firm;[15] these initiatives seem perhaps most important for maintaining social license or brand for Credit Suisse, not necessarily signals about a coming new phase of conservation capitalism.

And so, is biodiversity conservation financializing? There are clear signs that conservation is financializing in relationships and discourse, but the extent to which these are influencing the subjectivities, values and practices of biodiversity conservation is less clear. Further study will help us build clarity on the processes and stakes at the intersection of finance and conservation; I have laid out five interconnected research questions towards this end. Answering these questions cannot be achieved by reading documents or websites alone (although these can be illuminating); they demand systematic empirical research within organizations over time, concerted examination of funding patterns, close mappings of changing conservation priorities at multiple scales, developing a detailed understanding of any new risk-assessment or accumulation-oriented financial tools or vehicles. Based on research so far, however, there is good reason to be sceptical of claims that interest-bearing capital is racing to biodiversity conservation.

Given the financialization of supposedly everything else on the planet, why not 'securitizing nature to save it'? As I argue elsewhere (Dempsey and Suarez 2016), capital is not flowing to so-called 'conservation assets' because conservation is not an attractive investment; there is little evidence of a secure revenue stream that can repay debts or profit – the holy grail of securitization (see Leyshon and Thrift 2007). It does not seem to be the case that 'capital seeks to capitalize everything and everybody' (O'Connor, cited in Buscher and Fletcher 2015, p. 273), unless there is a 'clear and defined income stream' (Leyshon and Thrift 2007, p. 100) that has a chance of returning to investors. And it has not been easy to make an income stream from conservation. As the

NatureVest (2014, p. 12) report plainly states, conservation investments are much 'less competitive compared to competing market opportunities'. Returns from forest product commodities or agricultural commodities that cause forest conversion remain high. Returns from conservation remain low (or are an overall cost, as is most federal and provincial conservation). There is not widespread demand for carbon credits produced out of standing, biodiverse tropical forests, once thought to be the cornerstone of the coming green economy (see GCP et al. 2014). Governments are not rushing to develop policies that internalize externalities; neoliberal natures are standing off against neoliberal governments that are not very interested in the sixth extinction. And so my ambivalence about the financialization of biodiversity conservation deepens: I am not sure whether to pump my fist at its marginality or to crawl further into despair.

Conclusion: Critical Scholarship in the Sixth Extinction

What does it mean to be critical in these uncertain, dark ecological times, in a time when even the most status quo approaches to reducing biodiversity loss remain marginal? In her wonderful essay 'Untimeliness and Punctuality: Critical Theory in Dark Times', Wendy Brown (2009) reminds us that being critical necessarily means adherence to 'untimeliness'. Critical work, Brown suggests, ought to 'contest settled accounts of what time it is, what the times are, and what political tempo and temporality we should hew in political life' (p. 4). Critical theory in dark times must be a 'bid to reset time' (p. 4); it cannot be constrained by what is practical, pragmatic or politically strategic. Critiques of the neoliberalization of conservation achieve precisely this, refusing to be taken in by the argument that engagement with corporate and financial capital is the only way to save nature, drawing attention to how these approaches risk deepening the stranglehold capitalist logics and processes have in the production of nature. Scholars and activists critical of the financialization of nature offer us a different sense of the times and a different sense of time, insisting, as Brown (2009) would say, 'on alternative possibilities and perspectives in a seemingly closed political and epistemological universe' (p. 14).

Yet Brown states that critical theory in dark times must also exercise a profound reading of the times; it 'must grasp the age' (p. 14). And so, what if our times are not an age of great nature sale? How might we contend with *a lack of* interest by finance capital in conservation? Financial firms profiting off conservation at significant scale would at least suggest that biodiversity has value, politically and economically, on a planet where value is (if repulsively) tethered to capitalist value. I am not suggesting that capitalists are needed to save the planet, nor am I sidelining the work of social movements, Indigenous Peoples, farmers, community groups and NGOs in cultivating and protecting biodiversity. Yet grasping the age means looking straight into the abyss: we are living on a planet that is less lively, less bio-culturally diverse by the year, an earth, as Donna Haraway (2015) writes, 'full of refugees, human and not, without refuge' (p. 160). And the reality is, even the most palatable, status quo–affirming, neoliberal strategies to scale biodiversity conservation up – strategies emerging from places of significant cultural, political and economic power – remain marginal. To me, quoting Brown (2009, p. 10), 'the weight of the present' feels 'very heavy: all mass, no velocity', we are living with the weightiness of more non-human refugees, combined with the incredibly slow pace of action to safeguard or create refugia.

With our attention fixed on the *slowness* of state and capital response to the sixth extinction, one sees not the hyper-ability and mobility of market society in the face of crisis (i.e., the ability for capital to 'accumulate by conservation' in Buscher and Fletcher's 2015 terminology), but rather how embedded 'Cheap Natures' are to capitalist social relations. 'Cheap Nature' is a term of Jason Moore's (2015), referring to the enormous amount of unpaid, appropriated work non-human bodies and processes contribute to the production of capitalist value; Moore argues that cheap natures are a 'fundamental condition of capitalist accumulation' (Moore 2015, p. 2). If ending market failure (too many externalities, too much cheap nature) is the premise of many neoliberal environmental policies, then the underwhelming interest of finance capital in conservation provides evidence that cheap nature is enormously difficult to dislodge, *even by status quo political economic logic, actors and practices.*

Looking carefully into this abyss of '*non*-accumulation by conservation' could open up practical, even pragmatic, conversations about the *impossibility* of eliminating cheap nature within capitalist social relations – maybe even conversations between critics and those at the forefront of financializing biodiversity conservation. This non-accumulation by conservation vantage point may even open different analytical windows onto attempts to internalize externalities via economization, marketization or financialization. These strategies, viewed sceptically by critics (including myself) as new ways to profit from crisis, might be viewed as strategies to be appropriated by radical movements in the service of building human and non-human refugias.

Notes

1   The conference's invitation of Bellamy – who had been identified by George Monbiot as a leading climate-change denier (Monbiot 2009) – incited controversy between organizers and conservationists.

2   Accumulation by conservation is defined as 'a mode of accumulation that takes the negative environmental contradictions of contemporary capitalism as its departure for a newfound "sustainable" model of accumulation for the future' (Buscher and Fletcher 2015, p. 273).

3   Video is available https://vimeo.com/43398910, last accessed 20 Feb. 2017.

4   Cassimon et al. (2011) provide more concrete evidence on debt for nature: between 1987 and 1997, debt-for-nature swaps accounted for US$134 million of developing country debt, with the number of swaps declining in the mid-1990s due to the overall improving debt situation in countries like Mexico and Brazil. They also note that the main participant in swaps is the US, which between 2000 and 2007 participated in thirteen debt-for-nature swaps.

5   Global flow of finance is estimated at US$51.5–53.2 billion in 2010, with about US$40 billion in domestic funds, agricultural subsidies with conservation outcomes, and aid funding (Parker et al. 2012).

6   PES refer in general to monetary incentives given to landowners (or tenure holders) to manage their land for a particular ecosystem service (say carbon sequestration, or water filtration). Analysts have begun to argue that many if not most PES schemes fail to perform like markets and instead operate more like incentives (Boisvert et al. 2013; Fletcher and Breitling 2012; Muradian et al. 2012; Pirard and Lapeyre 2014; Pirard 2012).

7   Other TNC executives worked previously as investment bankers with Standard Chartered and Goldman Sachs and as capital market and corporate strategists at consulting giant McKinsey and Company. There is corporate influence too, with senior staff with former appointments at BP, Cirque de Soleil, Nike, Unilever and Staples.

8   Elites have long been at the centre of conservation, advocating for the first National Parks and starting the first conservation NGOs (Brockington and Duffy 2010). It is important to note that conservation comprises highly diverse elements; there are many organizations, movements and communities involved in conservation throughout the globe. Here I am speaking about international conservation conducted by large transnational organizations like Conservation International and The Nature Conservancy.

9   Impact investing is now a common term used to describe investments that aim to generate social, environmental and financial returns.

10   JP Morgan's head of commercial banking Doug Petno is the vice chair of the advisory board to NatureVest.

11   Other critical investigations include Adams and Redford (2010); McAfee and Shapiro (2010); Muradian et al. (2010); Kosoy and Corbera (2010); and Robertson (2011).

12 The Nature Conservation Note recently won the 2015 Environmental Finance award (Environmental Finance 2015).

13 A side note: the first credit default swap took place when JP Morgan agreed to extend credit to Exxon for its liability from the Valdez spill. To get this off its books, JP sold the debt to another firm in the first CDS (Tett 2010).

14 As of January 2016, eighty-one financial institutions have signed onto the Equator Principles. These principles are based on the International Finance Corporation's standards created in 2006 (the IFC is the private sector arm of the World Bank, investing in for-profit and commercial projects).

15 NatureVest appears in JP Morgan's website under Corporate Responsibility (JP Morgan and Chase 2016).

References

Adams, W.M., and K.H. Redford. 2010. Ecosystem services and conservation: a reply to Skroch and Lopez-Hoffman. *Conservation Biology* 24 (1):328–9.

Althelia. 2016. Althelia ecosphere. Viewed 26 January 2016, www.althelia.com.

Arrighi, G. 1994. *The Long Twentieth Century: Money, Power, and the Origins of Our Times*. London: Verso.

Barnaud, C., and M. Antona. 2014. Deconstructing ecosystem services: Uncertainties and controversies around a socially constructed concept. *Geoforum* 56:113–23.

Bishop, J., S. Kapila, F. Hicks, P. Mitchell and F. Vorhies. 2008. *Building Biodiversity Business*. Viewed 27 January 2016, https://portals.iucn.org/library/efiles/documents/2008-002.pdf.

Blackburn, R. 2006. Finance and the fourth dimension. *New Left Review* 39:39–70.

Boisvert, V., P. Méral and G. Froger. 2013. Market-based instruments for ecosystem services: Institutional innovation or renovation? *Society & Natural Resources* 26 (10):1122–36.

Bracking, S. 2012. How do investors value environmental harm/care? Private equity funds, development finance institutions and the partial financialization of nature-based industries. *Development and Change* 42 (1):271–93.

Brockington, D., and R. Duffy. 2010. Capitalism and conservation: The production and reproduction of biodiversity conservation. *Antipode* 42 (3):469–84.

Brown, W. 2009. *Edgework: Critical Essays on Knowledge and Politics*. Princeton: Princeton University Press.

Büscher, B. 2014. Nature on the move. In *Nature Inc: Environmental Conservation in the Neoliberal Age*, eds B. Büscher, W. Dressler and R. Fletcher. Tucson, AZ: University of Arizona Press,.

Büscher, B., and R. Fletcher. 2015. Accumulation by conservation. *New Political Economy* 20 (2):273–98.

Cassimon, D., M. Prowse, and D. Essers. 2011. The pitfalls and potential of debt-for-nature swaps: A US-Indonesian case study. *Global Environmental Change* 21:93–102.

Çalışkan, K., and M. Callon. 2009. Economization, part 1: Shifting attention from the economy towards processes of economization. *Economy and Society* 38 (3):369–98.

Ceballos, G., P.R. Ehrlich, A.D. Barnosky, A. García, R.M Pringle and T.M. Palmer. 2015. Accelerated modern human-induced species losses: Entering the sixth mass extinction. *Science Advances* 1(5).

Christophers, B. 2015. The limits to financialization. *Dialogues in Human Geography* 5(2):183–200.

Clark, S. 2007. *A Field Guide to Conservation Finance*. Washington, DC: Island Press.

Conservation Finance 2016. 2016. Conference: Conservation finance. New York: Credit Suisse. Viewed 27 January 2016, www.conservationfinance.ch/resources.

Credit Suisse, World Wildlife Fund and McKinsey & Company. 2014. *Conservation Finance: Moving Beyond Donor Funding Toward an Investor-driven Approach*. Viewed 26 January 2016, https://www.cbd.int/financial/privatesector/g-private-wwf.pdf

Dempsey, J. 2013. Biodiversity loss as material risk. *Geoforum* 43:41–51.

Dempsey, J. 2015. Ecosystem services. In *The International Encyclopedia of Geography: People, the Earth, Environment, and Technology*. Chichester: Wiley-Blackwell.

Dempsey, J. 2016. *Enterprising Nature*. Chichester: Wiley-Blackwell.

Dempsey, J., and M.M. Robertson. 2012. Ecosystem services: Tensions, impurities, and points of engagement within neoliberalism. *Progress in Human Geography* 36 (6):758–79.

Dempsey, J., and D. Suarez 2016. Arrested development? The promises and paradoxes of 'selling nature to save it'. *Annals of the American Association of Geographers* 106 (3):653–671.

Dirzo, R., H.S. Young, M. Galetti, G. Ceballos, N.J. Isaac and B. Collen. 2014. Defaunation in the anthropocene. *Science* 345:401–6.

Environmental Finance. 2015. Sustainable forestry: Credit Suisse/Althelia Ecosphere's Nature Conservation Notes. Viewed 27 January 2016, https://www.environmental-finance.com/content/deals-of-the-year/sustainable-forestry-credit-suisse-althelia-ecosphere.html.

Fletcher, R., and J. Breitling. 2012. Market mechanism or subsidy in disguise? Governing payment for environmental services in Costa Rica. *Geoforum* 43 (3):402–11.

Food and Agriculture Organization. 2010. *Second Report on the State of the World's Plant Genetic Resources for Food and Agriculture*. Rome: Food and Agriculture Organization.

Friends of the Earth. 2014. *Financialization of Nature*. Viewed on 27 January 2016, http://www.foei.org/resources/publications/publications-by-subject/forests-and-biodiversity-publications/financialization-of-nature.

GCP, IPAM, FFI, UNEP FI and UNORCID. 2014. Stimulating interim demand for REDD+ emission reductions: The need for a strategic intervention from 2015 to 2020. Global Canopy Programme, Oxford; the Amazon Environmental

Research Institute, Brasília, Brazil; Fauna & Flora International, Cambridge, UK; UNEP Finance Initiative, Geneva, Switzerland; and United Nations Office for REDD+ Coordination in Indonesia, Indonesia.

Gómez-Baggethun, E., R. de Groot, P.L. Lomas and C. Montes. 2010. The history of ecosystem services in economic theory and practice: From early notions to markets and payment schemes. *Ecological Economics* 69 (6):1209–18.

Hall, S. 2012. Geographies of money and finance II: Financialization and financial subjects'. *Progress in Human Geography* 36 (3):403–11.

Harvey, D. 2005. *A Brief History of Neoliberalism*. Oxford: Oxford University Press.

Huwyler, B.F., J. Kaeppeli, K. Serafimova and E. Swanson. 2014. *Making Conservation Finance Investable*. Viewed 27 January 2016, http://www.ssireview. org/up_for_debate/article/making_conservation_finance_investable.

International Finance Corporation. 2012. Performance standards on environmental and social sustainability. Viewed on 25 January 2016, http://www.ifc. org/wps/wcm/connect/115482804a0255db96fbffd1a5d13d27/ PS_English_2012_Full-Document.pdf?MOD=AJPERES.

International Indigenous Peoples Movement for Self Determination and Liberation. 2013. Declaration of indigenous peoples on the World Trade Organization (WTO). Viewed 27 January 2016, http://ipmsdl.wordpress. com/2013/12/08/declaration-of-indigenous-peoples-on-the-world-trade-organization-wto.

Johnson, L. 2013. Index insurance and the articulation of risk-bearing. *Environment and Planning A* 45:2663–81.

JP Morgan Chase & Co. 2016. *Driving Sustainability Through Business: Environmental and Social Risk Management*. Viewed 25 January 2016, http://www.jpmorganchase.com/corporate/Corporate-Responsibility/ driving_sustainability_through_business.htm.

Knudson, C. Forthcoming. The insurance trap: Banana farming in Dominica after Hurricane Hugo. In D. Barker, D. McGregor, K. Rhiney and T. Edwards, eds. *The Caribbean Region: Adaptation and Resilience to Global Change*. Kingston, Jamaica: University of the West Indies Press.

Kolbert, K. 2009. The sixth extinction. *New Yorker* 25 May 2009. http://www. newyorker.com/magazine/2009/05/25/the-sixth-extinction.

Kosoy, N., and E. Corbera. 2010. Payments for ecosystem services as commodity fetishism. *Ecological Economics* 69:1228–36.

Krippner, G.R. 2005. The financialization of the American economy. *Socio-Economic Review* 3 (2):173–208.

Kull, C.A., X. Arnauld de Sartre and M. Castro-Larrañaga. 2015. The political ecology of ecosystem services. *Geoforum* 61:122–34.

Leyshon, A., and N. Thrift. 2007. The capitalization of almost everything: The future of finance and capitalism. *Theory, Culture & Society* 24:97–115.

MacDonald, K.I. 2010. The devil is in the (bio)diversity: Private sector 'engagement' and the restructuring of biodiversity conservation. *Antipode* 42 (3):513–50.

MacKenzie, D. 2004. The big, bad wolf and the rational market: portfolio insurance, the 1987 crash and the performativity of economics. *Economy and Society* 33 (3):303–34.

MacKinnnon, J.B. 2010. A 10 percent world. *The Walrus.* Viewed 27 January 2016, https://thewalrus.ca/a-10-percent-world.

Mandel, J., J. Donlan J. and Armstrong. 2010. A derivative approach to endangered species conservation. *Frontiers in Ecology and the Environment* 8 (1):44–9.

Martin, R. 2002. *Financialization of Daily Life.* Philadelphia: Temple University Press.

McAfee, K. 1999. Selling nature to save it? Biodiversity and green developmentalism. *Environment and Planning* 17:133–54.

McAfee, K., and E.N. Shapiro. 2010. Payments for ecosystem services in Mexico: Nature, neoliberalism, social movements, and the state. *Annals of the Association of American Geographers* 100 (3):579–99.

McGregor, A., S. Weaver, E. Challies, P. Howson, R. Astuti and B. Haalboom. 2014. Practical critique: Bridging the gap between critical and practice-oriented REDD+ research communities. *Asia Pacific Viewpoint* 55 (3):277–91.

Monbiot, G. 2009. Monbiot's royal flush: Top ten climate deniers. *Guardian,* 9 March. Viewed 27 January 2016, http://www.guardian.co.uk/environment/georgemonbiot/2009/mar/06/climate-change-deniers-top-10.

Monboit, G. 2014. The pricing of everything. Viewed 27 January 2016, http://www.monbiot.com/2014/07/24/the-pricing-of-everything.

Moore, J.W. 2015. *Capitalism in the Web of Life.* London: Verso.

Moringa Fund. 2016. Moringa Fund: About us. Viewed 27 January 2016, http://www.moringapartnership.com/web.php/16/en/about-us/organisation?

Muradian, R., et al. 2010. Reconciling theory and practice: an alternative conceptual framework for understanding payments for environmental services. *Ecological Economics* 69 (6):1202–8.

Muradian, R., et al. 2013. Payments for ecosystem services and the fatal attraction of win-win solutions. *Conservation Letters* 6 (4):274–9.

NatureVest. 2016. NatureVest: Our projects. Viewed 27 January 2016, http://www.naturevesttnc.org/our-projects.

Norgaard, R.B. 2010. Ecosystem services: from eye-opening metaphor to complexity blinder. *Ecological Economics* 69 (6):1219–27.

Parker, C., M. Cranford, N. Oakes and M. Leggett. 2012. *The Little Biodiversity Finance Book: A Guide to Proactive Investment in Natural Capital (PINC).* Available at: https://www.cbd.int/financial/hlp/doc/literature/LittleBiodiversity FinanceBook_3rd%20edition.pdf.

Partners in Sustainable Land Management. 2016. SLM: Home. Viewed 25 January 2016, http://slmpartners.com.

Pirard, R. 2012. Market-based instruments for biodiversity and ecosystem services: A lexicon. *Environmental Science & Policy* 19–20:59–68.

Pirard, R., and R. Lapeyre. 2014. Classifying market-based instruments for ecosystem services: A guide to the literature jungle. *Ecosystem Services* 9:106–14.

Prudham, S. 2003. Taming trees: Capital, science, and nature in Pacific Slope tree improvement. *Annals of the Association of American Geographers* 93 (3):636–56.

Robertson, M. 2012. Measurement and alienation: making a world of ecosystem services. *Transactions of the Institute of British Geographers* 37 (3):386–401.

Smith, N. 2007. Nature as accumulation strategy. *Socialist Register* 43:16.

Sullivan, S. 2009. Green capitalism, and the cultural poverty of constructing nature as service provider. *Radical Anthropology* 3:18–27.

Sullivan, S. 2010. 'Ecosystem service commodities': a new imperial ecology? Implications for animist immanent ecologies, with Deleuze and Guattari. *New Formations* 69 (1):111–28.

Sullivan, S. 2012. *Financialisation, Biodiversity Conservation and Equity: Some Currents and Concerns.* Third World Network, Penang, Malaysia. Viewed 27 January 2016, http://www.twn.my/title/end/pdf/end16.pdf.

Sullivan, S. 2013. Banking nature? The spectacular financialisation of environmental conservation. *Antipode* 45 (1):198–217.

Sullivan, S. 2014. The natural capital myth; or will accounting save the world? Preliminary thoughts on nature, finance and values. *Leverhulme Centre for the Study of Value, Working Paper Series No. 3.* Viewed 27 January 2016, http://thestudyofvalue.org/wp-content/uploads/2013/11/WP3-Sullivan-2014-Natural-Capital-Myth.pdf.

Tett, G. 2010. *Fool's Gold.* Toronto, ON: Simon and Schuster.

The Nature Conservancy. 2016. Working with Companies: SwissRe. Viewed 27 January 2016, http://www.nature.org/about-us/working-with-companies/companies-we-work-with/swiss-re.xml.

United Nations Environment Programme Financial Initiative and Global Canopy Programme. 2012. *The Declaration.* Viewed 27 January, http://www.naturalcapitaldeclaration.org/the-declaration.

Waage, S., and C. Kester. 2014. *Making the Invisible Visible: Analytical Tools for Assessing Business Impacts & Dependencies upon Ecosystem Services.* Viewed 27 January 2016, http://www.bsr.org/reports/BSR_Analytical_Tools_for_Ecosystem_Services_2014.pdf.

Waage, S., E. Steward and K. Armstrong. 2008. *Measuring Corporate Impact on Ecosystems: A Comprehensive Review of New Tools.* Business for Social Responsibility Network, viewed 29 January 2016, http://www.bsr.org/reports/BSR_EMI_Tools_Application_Summary.pdf.

World Economic Forum. 2015. *Global Risks 2015*, Viewed 27 January 2016, http://www3.weforum.org/docs/WEF_Global_Risks_2015_Report15.pdf.

World People's Conference on Climate Change and the Rights of Mother Earth. 2010. Peoples Agreement. Viewed 27 January 2016, https://pwccc.wordpress.com/support.

9

Financialization of Singaporean Banks and the Production of Variegated Financial Capitalism

Karen P.Y. Lai and Joseph A. Daniels

Size matters in international banking [… The local banks] need to grow large enough to enjoy the economics of scale, and to have the reach and resilience to go regional, and eventually make a mark in global markets. This is why MAS [Monetary Authority of Singapore] has encouraged local banks to consider mergers. (MAS, 1998)

Introduction

The 2008 global financial crisis has refocused popular and scholarly attention on the vital roles of banks in national and global economies. As the credit crunch unfolded, the collapse, bail-outs and mergers of banks in the USA, UK and Europe prompted reflections on whether the ever-growing clout and economic impact of banks deemed 'too big to fail' had become detrimental to national economies and global finance. Almost two decades ago, a different kind of debate took place in Singapore after the 1997 Asian financial crisis, as banks in Singapore were challenged to grow bigger in order to extend their extraterritorial reach and secure long-term competitiveness. The responses by the

Money and Finance After the Crisis: Critical Thinking for Uncertain Times, First Edition. Edited by Brett Christophers, Andrew Leyshon and Geoff Mann. © 2017 John Wiley & Sons Ltd. Published 2017 by John Wiley & Sons Ltd.

Singaporean government to the Asian financial crisis reconfigured the structure of the Singaporean banking system in ways that shape the impacts of and responses to the 2008 global financial crisis. In broadening the temporal scope of the analysis to the Asian financial crisis, we avoid 'fetishizing' the 'global' financial crisis as universal in its impact across space and time; it is not the same historical marker in Singapore as it is in the North Atlantic. The global financial crisis offers a moment to consider how we deal with change but also continuity when thinking through finance in the post-crisis period, particularly as the trans-formations following the Asian financial crisis, which we illustrate in the following sections, created certain vulnerabilities that became apparent later during the 2008 global financial crisis.

The 1997 Asian financial crisis shook the banking systems of Asian economies and crippled manufacturing sectors that had been reliant on cheap capital and foreign inputs for production. It also shook the confidence of foreign investors and domestic enterprises, and compelled businesses and policymakers to rethink their strategies in Asia (Arndt and Hall 1999; Poon and Thompson 2001). Although Singapore was among those least affected in Asia, it did not escape the contagion effects of the Asian financial crisis (Henderson 1999). Weak incomes, restricted bank liquidity, labour retrenchment and reduced regional trade culminated in Singapore's second economic recession since independence. Economic stimulation measures included higher tax rebates, increased public expenditure and government-led investment in new growth sectors such as advanced engineering, chemicals, media and communications (Lai 2013a). The banking sector was also earmarked as part of a strategy to secure the longer-term competitiveness of Singapore's economy (Committee on Singapore's Competitiveness 1998). One objective was to develop large domestic banks that would have enlarged business networks into regional and global markets in order to support the regionalization efforts of other Singaporean firms during that period. This resonates with the long-term aspiration of the Monetary Authority of Singapore (MAS), Singapore's central bank and financial regulator, to develop Singapore as a preeminent international financial centre (IFC) with local banks to match that status. While Singapore's export-led manufacturing growth strategy has been well documented (see for example Rodan 1989; Perry et al. 1997; Low 2001), the finance sector has also featured prominently in Singapore's economic development plan since the 1980s (Hamilton-Hart 2002; Tan and Lim 2007). For Singapore after the Asian financial crisis, a key challenge was to develop the 'right' kind of banking institutions deemed necessary to secure IFC status. This became known as the 'Big Bang' reform (Ngiam 2011: 188).[1]

Before 1998 there were six major local banks: the government-linked DBS Bank (formerly Development Bank of Singapore), Keppel TatLee, and Post Office Savings Bank (POSB owned by the Ministry of Finance), alongside family held banks namely United Overseas Bank (UOB), Oversea-Chinese Banking Corporation Limited (OCBC) and Overseas Union Bank (OUB). The 'Big Bang' started in 1998 with the sale of POSB to DBS by the Singapore government, making DBS the largest bank in Singapore. After a failed hostile takeover, OUB merged with UOB, and OCBC acquired the remaining Keppel TatLee (Figure 9.1). This process of capital centralization was propelled by a government responding to international trends in global finance and the perceived inadequacies of local firms and national developmental strategies (Cook 2008). The evolution of Singapore's banking regulatory environment and changes in business strategies of Singaporean banks post-crisis present a useful lens through which we could analyse the divergent unfolding of financialized capitalist change. The financialization of Singaporean banks (here understood broadly as their transformation into more diversified financial services corporations) from the early 2000s has important implications for their changing business orientation towards securitization as well as increased emphasis on consumer markets for fee-based

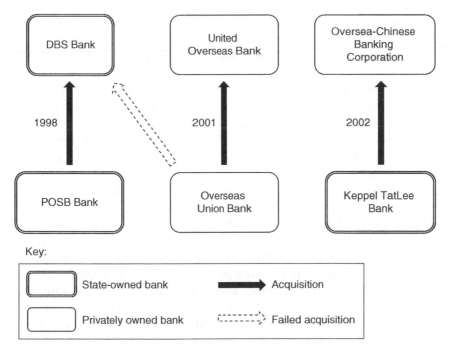

Figure 9.1 Mergers of local banks in Singapore.

activities; these would have significant impacts on the banks, particularly DBS, during the 2008 global financial crisis (a point to which we return in the Conclusion).

In this chapter we analyse the role of the state vis-à-vis local banking firms during a period of industrial change and consider some implications regarding the nature of capitalist change. In doing so, we make two points. First, the transformation of local banks into financial services corporations demonstrates the vital and strategic role of the state in the financialization of firms. Rather than treating the state as a distant or reactionary agent providing the background of deregulation for greater market efficiency, we focus on state–firm relations in mobilizing financialization processes and their developmental outcomes. Second, what might be seen as a convergence in governance and organizational structures of local banks in this financialization process actually stems from strikingly different relationships with the state, such that the transition to financialized modes of production is more contested than cohesive; what we call variegated financial capitalism. This draws from the notion of variegated capitalism, i.e., capitalism as a singular, but polymorphic and uneven, process. Instead of models based on national archetypes (Peck and Theodore 2007; Dixon 2011), and adding just another 'variety' in the Asian portfolio of capitalisms (see Yeung 2012), we unpack the variegated financialization of banking firms in Singapore by moving beyond the nation as analytical container to uncover the multiscalarity and complex unfolding of capitalist processes. We seek to bring together financialization and varieties of capitalism (VOC) literatures in order to understand the state and its role in shaping capitalist geographies and trajectories for firms and economies. As finance transcends its traditional intermediary role to become an increasingly self-standing growth industry primarily geared to the production of liquidity, it has also become an increasingly important component of business strategies and operational logics, bringing about significant structural transformation in contemporary capitalism (Engelen and Konings 2010; Christophers 2013; Hall 2013).

Singapore provides an opportunity to examine the dynamics of capitalist change during a period of financialization, i.e., the transformation of local firms from *banking* firms to *financial services* corporations as the result of state policy. In transforming local banks into 'financial' firms through strategic policy implementation, the state hoped to create demand for more sophisticated financial markets well beyond the bounds of the target banks. In this context, banking refers to the traditional activities of loan mediation, while financialization is a shift towards financial logics, including but not limited to the adoption of a shareholder conception of control, focus upon financial markets and products, and the enlargement of non-bank financial investments in insurance and similar functions (Dore 2008; Erturk et al. 2008; Froud et al. 2006; Williams 2000). The two case studies – DBS and OCBC – are firms whose strengths are deemed necessary by the state for developing Singapore into a robust and successful IFC, but who have responded differently to 'Big Bang' reforms. Therefore, while regulatory and industry changes may present a broad picture of financialized firm behaviour, the actual mode of financialization is contingent on institutional histories, social networks and the existing geographical distribution of business operations. Our approach takes up Dixon's (2011) challenge to better incorporate the firm into an analysis of variegated capitalism, especially in placing banks as active agents of capitalist production rather than mere intermediaries (Christophers 2013). Scrutinizing the state–firm nexus of Singaporean banks' financialization, the findings deepen our understanding of financialized modes of production, state–firm relations in capitalism and variegated financial capitalism.

The next section consists of a critical review of the literature on varieties of/variegated capitalism and financialization. The empirical discussion adopts a multiscalar approach in analysing how changes in global financial regulatory frameworks, national regulatory regimes, regional aspirations and local firm behaviour are entangled in the production of economic strategies and financial expansion. The case studies of DBS and OCBC are then used to explicate how state policies shape the business strategies and geographical footprints of these firms as they seek to become larger regional players but in accordance with their different firm histories and cultures. In conclusion, we reflect on how combining insights from variegated capitalism and financialization could deepen our understanding of contemporary capitalism.

Variegated Capitalism, Variegated Financialization

Variety begets variegation: Too many models to count?

Since the late 1990s, the VOC school has established itself as a critical counter-current to the universal market rationality of globalization narratives, pointing to resilient differences in national capitalist regimes. The persistence of national varieties of capitalism is explained by the complex embedding of firms' and other actors' strategic behaviour in a range of institutional environments, the establishment of institutionally mediated forms of comparative advantages and the tendency for non-convergent, path dependent evolution in national regimes (Campbell 2004). Most notably, Hall and Soskice (2001) claim capitalism has developed two distinct models – the American-styled capitalism of liberal market economies (LMEs) and the German model of coordinated market economies (CMEs).

These two models describe some viable alternatives to the free market trajectory, and help explain the differentiated character of actually existing capitalist formations. Unfortunately, this 'methodological nationalism' has significant limitations (Peck and Theodore 2007). First is the issue of accommodating or explaining 'exceptions'. Second is the a priori assumption of national coherence in the face of regional and local variation. Finally, the model's focus on (national) institutionalism can elide the central object of analysis: capitalism, which is multiscalar and constituted by multiple actors (Howell 2003; Crouch 2005; Deeg and Jackson 2007).

Despite the VOC assumption of capitalist diversity, there has been limited theoretical scope for explaining institutional transformation and hybrid configurations, and the analytical monopoly of the nation-state remains conventional practice.[2] There has been a flurry of models that attempt to capture the wider variation in capitalism, producing ever more distinct 'national' models such as market-based, Asian, Continental-European, Social-Democratic and Mediterranean (Amble 2003). Rather than providing greater clarity, however, this proliferation overstretches the VOC idea by trying to accommodate too many 'exceptions'. When exceptions become the norm it is difficult to develop productive theory, and requires new theoretical intervention. The focus on national differences also overlooks the significance of networks and practices in market formation and governance, constituted by individuals, firms and other institutions operating both within and beyond the national scale.

Consequently, Peck and Theodore (2007: 733) argue for 'variegated capitalism':

> a shift away from the varieties-style reification and classification of economic-geographical difference, in favour of a more expansive concern with the combined and uneven development of 'always embedded' capitalism, and the polymorphic interdependence of its constitutive regimes.

This concept of 'variegated capitalism' attempts to deepen theoretical understanding of variety to incorporate the complex interrelations, differences and similarities that exist in and between different spaces of capitalism. Instead of discrete models of simple convergence or divergence, capitalist systems should be analysed as *layered and hybrid* forms of capitalist variation in geographically and historically contingent sets of circumstances.

The variegated capitalism framework provides a theoretical release from any singular model, in favour of understanding capitalism in its uneven and recombinant forms through the complexity of state–capital relations. Instead of focusing on 'multiple capitalisms', this approach views 'capitalism in the singular, but more importantly as a dynamic polymorphic process whose development is uneven and "variegated"' (Dixon 2011: 197). Rather than ascribing a model of capitalism from the top down, we view Singapore in relational and conjunctural terms, rather than terrain of typicality or exception. We therefore situate capitalist development and change through the accounts of diverse actors as negotiations unfold amidst different interpretations of what constitutes appropriate or desirable market structures and practices in the pursuit of financialized growth. This means that an account of the financialization of Singaporean banks need not be one that relies upon a national 'model', but rather one that explains how common broader trends and local institutions interact to produce varied capitalist practices and outcomes.

Financialized capitalism

There now exists many definitions of financialization (see van der Zwan 2014). Krippner's (2005) much-cited definition identifies financialization as 'a pattern of accumulation in which profits accrue through financial channels rather than through trade and commodity production' in the national economy and large corporations of the USA (p. 174). Studies of financialization range from how the finance sector dominates national political economies (Blackburn 2006; Dore 2008), to how firm strategies and management are increasingly beholden to the logics of finance (Williams 2000; Krippner 2005; Froud et al. 2006; Ho 2009), and the ways in which households and individuals are tied into increasingly complicated relationships with the international financial system (Martin 2002; Langley 2008; Lai 2017).

Studies from the regulation school typically concern the nature of capitalism and how that has changed under conditions of financialization. In particular, scholars debate the apparent shift from industrial to financial capitalism. Political economists and Marxian theorists such as Arrighi (2003) suggest that financialization is a capitalist response to the crisis of spatial-state formation in the pursuit of profitability and hegemony, with multiple rounds of deregulation and government support fuelling the growth of the finance industries (Helleiner 1995) and reifying the prevalence of financialization as techniques of governance (Harvey 2005; Dymski 2009; Krippner 2012). The increasing pressure of shareholders for growing corporate value has induced management to seek out wealth creation in non-traditional venues such as financial and property markets, rather than through production or innovation (see Froud et al. 2000; 2006, Clark et al. 2002; Engelen 2003).

Given the claims made for the pervasive power of finance across various segments of economy and society, the concept of financialization is becoming a metanarrative or new master concept to capture and explain the dynamics of financial change (Engelen and Konings 2010: 605). However, the financialization of different spaces, and how space, place and specific actors are mobilized in financialization processes are not always specified (Pike and Pollard 2010; French et al. 2011). Instead of a unified and hegemonic process, financialization is a much more spatially variegated phenomenon, which has important implications for understanding capitalism and changing modes of production. For the most part, the VOC literature has little to say about finance except for some basic distinction between bank-based (CME) and market-based (LME) forms of financing for production and firm activities. But as financial capitalism continues to transform the very nature of these capitalist processes and expand the economic diversity of institutions and financial agents, the distinction between the 'real economy' and 'financial sector' is increasingly problematic (Hall 2013). Studies of contemporary economic change therefore must consider how finance shapes the construction of socio-spatial dynamics in the capitalist system. Engelen and Konings (2010), for instance, employ the US and UK (LME), Germany and France (CME), and the Netherlands (hybrid) as examples to suggest differing modes of financialization; consensual, contested and compartmentalized respectively. Their analysis points to the power of financial agents in shaping the geometries of power within their respective financial regulatory environments for wealth creation, resulting in different capitalist growth dynamics.

The role of the state also bears rethinking in the context of financialization. A dominant strand in the literature positions financialization as a retreat of state functions, with increasing reliance on financial logics and market-based solutions for social welfare (Clark 1998; Martin 2002), and overall state 'decline' in the advance of neoliberalism (Duminel and Lévy 2004; Stiglitz 2003). More recently, scholars have suggested that far from 'state retreat', different forms of intervention have led to the production of the financialized economy (see Crouch 2011; Prasad 2012), particularly in terms of financial deregulation and institutional change, firm behaviour and everyday habits of saving and borrowing (Langley 2008; Krippner 2012). However, despite their concern with the role of regulation, these analyses still tend to emphasize market imperatives and neoliberal logics over state power and functions, and tend to dwell on state incapacities to resolve internal crises and emphasize deregulation for greater market efficiency. Such a perspective underplays the strategic ways in which the state actively mobilizes institutions and firms to adopt and enact financialization scripts for political-economic purposes, i.e. state-led financialization.

In response, we argue for a reconsideration of how states reconfigure their roles and reposition firms within the global financial system. This requires closer examination of state–firm relations. Following Dixon (2011), who advocates incorporating firm-level analysis and geographies of finance in understanding variegated capitalist development, we see a more robust variegated capitalism framework as infused with financialization logics. VOC approaches have largely confined finance to second-order functions subsumed into 'the market' rather than as a key actor in the production of capitalist processes and outcomes. By analysing the ways in which financialization is constructed through the state–firm nexus, we can develop a richer conceptualization of the state in financialization, and the implications for shifting configurations of capitalist growth.

State-driven Financialization: Global, National and Firm-based Strategies

This section traces the development of state-driven strategy in the growing importance of finance in Singapore's long-term economic growth, and subsequently the incorporation of financialized modes of production in firm organizational structures and corporate behaviour. The finance sector already featured prominently in Singapore's economic development plan prior to the Asian financial crisis. In 1985 Singapore faced its first recession since independence, and its first government deficit (due to depressed demand for manufactured goods and the petro-dollar debt crisis) (Tan 2005). A Sub Committee on Banking and Financial Services (SBFS) review identified two challenges: deregulation of financial markets and trends towards securitization – both deemed necessary to develop into a 'premier financial centre' (SBFS 1985: i). The report also called for the MAS to 'take on a more developmental role' like that of the Economic Development Board (EDB) (ibid.: iv) in order to boost the financial services industry and to contribute to the long-term economic growth of Singapore.[3] This marked a departure from Singapore's export-led manufacturing strategy since the 1970s (see Rodan 1989), and became the outline for policies linking forms of financialization – a greater role for capital markets as sites of capital growth – to the production of Singapore as a space (or spatial fix) for global capital.

The focus on banking and finance as a key pillar of growth reemerged in the 1998 report of the Sub-committee on Banking and Finance (SBF) following the Asian financial crisis (Committee on Singapore's Competitiveness 1998). This report focused exclusively on 'non-traditional' areas of finance that would enable Singapore to become a premier financial centre – including fund management, risk management, equity markets, debt insurance, corporate finance, insurance and reinsurance and cross-border banking. A national strategic plan known as 'Regionalization 2000' was launched to develop an 'external wing' to Singapore's economy, through overseas expansion of domestic firms and industrial parks in other Asian countries (Yeung 2000; 2004). The regionalization of domestic firms was not limited to manufacturing and government-linked corporations, but also involved Singaporean banks. The shift in spatial framing beyond the national space-economy was a necessary precondition for the financialization of Singaporean banks as the state viewed existing 'local' assets as being too limited to provide for the types of robust financial market participation it envisaged. The state's focus upon finance as the major site of Singapore's prosperity is therefore closely linked to the contours of its extraterritorial economy.

The entanglement of global and local imperatives can be seen in how compliance with and participation in international financial governance structures are fused with national developmental objectives in order to justify state policies. The liberalization programme started shortly after Singapore joined the Bank of International Settlements (BIS) in 1996, followed by membership on the Basel Committee on Banking Supervision (BCBS) in 2009. However, the work with the BIS and the BCBS was not just a transfer of the Basel policy model to Singapore; rather, it was a combination of 'demand-side needs, imperatives, and anxieties' and 'supply-sided inventiveness' characteristic of variegated processes (Peck and Theodore 2010: 171). While Singapore was reconfiguring its role on international regulatory platforms, it also had to find solutions for building more robust financial institutions in the wake of the Asian financial crisis.

A key policy shift was a more consultative 'risk-based' model of regulation that was in line with the Basel II requirements, rather than the previous 'one size fits all' supervisory approach (Ong 2004). This enabled individual firms to exercise greater freedom in new markets and sectors but also required them to adopt internal measures to comply with broader regulatory guidelines (such as capital ratio and reporting requirements, etc.) (Hamilton-Hart 2002). By shifting the rationale for regulatory changes outside the territory of Singapore, liberalization measures were justified by global regulatory standards. At the same time, banks in Singapore could consider which of the approaches best suit their respective institutions, taking into account the complexity and risk profile of their businesses (Ong 2004). By adopting liberalization policies and mobilizing international agreements, the state drew upon the discursive power of global regulatory compliance, while at the same time opening up new space for local firms to take responsibility for their internal control and risk management.

The impact of these 'global imperatives' was softened by a more 'consultative' and 'business friendly' style of supervision that was subsequently institutionalized through a new practice of consultation papers. The circulation of such consultation papers serves a valuable purpose, enabling the MAS and local finance industry to discuss proposed regulatory changes and fine-tune policies before final implementation. These consultative practices are integral to state–firm relations, illustrating Singapore's distinctive form of 'consultative authoritarianism' (Rodan 2012), in which professional and economic elites are co-opted into state institutions to enable the smooth operationalization of economic policies (Jayasuriya and Rodan 2007; Woo 2014). For instance, a structure of governing elites consisting of former politicians, civil servants and regulators circulating between state institutions and governing bodies of banks is common in Singapore as well as other Asian economies and encourages compliance with financial policies and developmental programmes (Hamilton-Hart 2002).

The transformation of the local banking industry into a robust globally oriented financial services sector is one such developmental goal vital to Singapore's long term competitiveness and economic success (Ngiam 2011). The restructuring of Singapore's banking system has unfolded in an environment of contradictory and emergent political-spatial imaginaries (see Jessop 2002). On one hand, there is recognizable neoliberalization, forcing local firms to upgrade (or upsize) and become more efficient by increasing foreign competition in local markets and shifting responsibility on to firms. On the other hand, there are spaces of 'consultation', a form of state-industry collaboration to shape the financial industry towards the features necessary for a sophisticated IFC, which enable flexibility and opportunities for stable state–firm relations. Rather than a sea change to a financialized mode of capitalist growth and firm behaviour, there are multiple eddies and currents operating in persuasive, tentative and sometimes contested ways. Not only are the geographies of capitalism not nearly as neat as the VOC literature would suggest, they are also increasingly shaped by financial logics.

The banking reforms of 1999–2004 had three main components: mergers, divestment requirements and changes in corporate governance.

The merger of six major local banks into three significantly increased deposit bases, deemed vital (by the state) to promote extra-territorial competitiveness. The enlarged banks were also supposed to expand their non-deposit taking business, transforming themselves from traditional banks to more complex financial institutions offering an extensive and sophisticated array of products and services to an expanded regional and global customer base (Cook 2008). In 2000, local banks were required to divest themselves of their non-financial businesses and unwind cross-shareholdings within three years. This not only complied with Basel requirements, but also allowed local banks to rebuild their financial position following the Asian financial crisis (Brown 2006). The divestment policy included four elements pertaining to firms' ownership, cross-shareholding, management and name sharing (Table 9.1).

The reforms were also concerned with corporate governance. In a speech to the Association of Banks in Singapore, then deputy Prime Minister Lee Hsien Loong (Lee 21 June 2000) contrasted the 'spry competitive banks' of the US with the German, Japanese and Korean models tied down by various cross-listing structures, seemingly encumbered in the context of 'an intensely competitive and dynamic [...] globalized industry'. This point was clearly directed at the family owned banks (OCBC, OUB and UOB). In 2001, OUB rebuffed a hostile takeover from

Table 9.1 Four elements of the MAS divestment policy

Regulatory Target	Requirement
Corporate entity/parent	• All financial activities must be held under a bank or financial holding company to be regulated by MAS.
Cross-shareholding	• Shareholders in the financial arm cannot be shareholders in non-financial arms, with a separate listing for the bank or financial holding company on the stock exchange. • Mutual shareholding between firms in the financial arm or between non-financial firms is not permitted. • Financial firms should not own shares in non-financial firms related to principal shareholder.
Management	• Management of non-financial firms and financial firms must be distinct.
Name sharing	• Financial firms and non-financial firms must not share the same name, for this could impinge on the reputation of the banks.

Source: Lee, 2000.

DBS, fearing loss of family control following such a merger. When OUB successfully merged with UOB (another family-founded bank) under more favourable terms, the deal was criticized as 'an attempt to keep family control intact without regard for shareholder value' (Low I. 2001). In this case, becoming corporations deemed fit for global competition clearly meant breaking away from family control and management.

The transition from banks to financial companies was driven by financialized conceptions of shareholder value and global competition. In the post-Basel II environment, local banks faced the challenge of becoming 'world class' banks, capable of competing with the likes of JP Morgan or

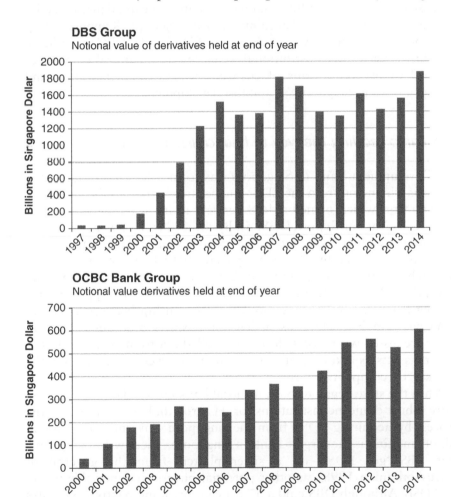

Figure 9.2 Dramatic growth in derivatives volume of DBS and OCBC banks. *Sources*: DBS annual reports 1997–2014; OCBC annual reports 2000–2014.

Citigroup, which meant changing not only their organizational structures and corporate governance but also reorienting their profit streams towards greater capital market participation and product innovation. Over the past decade both DBS and OCBC significantly expanded their investment banking and/or asset management businesses through the development and acquisition of new subsidiaries (DBS Securities 2001; OCBC Annual Report 2003). These were instrumental in driving trade in more complex financial instruments (e.g., derivatives), which formed an increasingly important part of bank business and profit (see Figure 9.2). This is a clear demonstration of financialization at work, i.e., the increasing importance of financial instruments and capital market activities to firms' operation and accumulation strategies (cf. Krippner 2005). In the next two sections, we use the cases of DBS and OCBC to explicate how firms' variegated institutional, economic and cultural embeddedness led to different forms of engagement with financialization and changing roles in Singapore's capitalist development.

DBS Bank: Leading the way to financialized growth

The Development Bank of Singapore was formed in 1968 to take over industrial financing from the EDB[4] and became banker to the Singapore Government, which owned a 46.8 per cent share in DBS at the end of 1968 (DBS Annual Report 1968: 5). By the 1980s, DBS began to move into commercial banking, and was particularly significant as a lender in the property market. In 1983, it adopted the 'DBS Bank' moniker as part of a rebranding from government-linked corporation to commercial enterprise (*Business Times* 5 April 1990). Despite this apparent distancing from its government roots, the relationship between DBS and MAS has remained close, as policies or business practices devised in either are tested with DBS before being rolled out to the wider industry. DBS thus acts as a key agent of state initiatives even if it engages in the 'market' on competitive terms.

When the state began to push for local bank consolidation to develop more robust domestic institutions for IFC growth, DBS spearheaded the process by acquiring POSB from the Singapore Government for S$1.6 billion, making it the largest bank in Singapore. There were complaints that DBS 'got POSB on a silver platter' from the government (Purushothaman 1998), as other potential buyers were not able to bid for POSB. Although DBS paid a 37 per cent premium relative to POSB's market valuation at that time, the takeover enabled DBS to go from having the third smallest deposit customer base to the largest. This dramatically altered the landscape of retail banking in Singapore and pressured smaller banks such as OCBC and OUB to seek out mergers themselves. When DBS heeded the government's 'Regionalization 2000' rhetoric, its foray into regional markets, particularly in Hong Kong and China, was quickly followed by the other banks (Phelps and Wu 2009; Chia 2011; Yeung 2000).

Specific policies in the banking reforms of 1998–2004 can be traced to experimental strategies enacted by DBS after the 1997 Asian financial crisis. To deal with losses from regional ventures, DBS hired foreign leaders for business reorganization, divested itself of non-financial assets and reorganized them into financial holding companies (Daniel 2001). MAS would later adopt these strategies as best practice in its 1999 liberalization scheme and 2000 divestment requirements. Such changes were deemed necessary for Singapore banks' expansion overseas and refocus on core assets such as financial services, turning DBS into a 'financial supermarket rather than a messy conglomerate' (Raj 1999). The shift toward financial investments and derivative trade has been instrumental to DBS's success; profits from financial investments nearly doubled between 1998 and 2011, and derivatives trade grew from approximately S$31 million to S$1.6 trillion. This financialized business mode of growth, made possible by reorganization of the holding company, was deemed instrumental for an IFC with greater market depth and sophistication, and was later taken up as MAS policy in banking reforms.

The organization of Singapore's banks was directly affected by the ways in which the state and DBS mutually constructed acceptable forms of operation, leading to the production of new, larger financial services firms, rather than just banks. The Development Bank of Singapore became 'DBS', 'a leading financial services group in Asia [...] headquartered in Singapore [...] a bank born and bred in Asia'.[5] Banks were beginning to encompass all forms of intermediation that go well beyond loans, mediation and beyond Singapore's national borders. In its reorganization of national economic strategies, DBS was mobilized as an active agent of the state. In 2001, at the suggestion of the Singaporean government (as DBS's major shareholder), it acquired Dao Heng Bank in Hong Kong. Although the bank's shareholder value took a large hit due to the premium paid for takeover, an industry analyst remarked:

> I am not sure this deal is necessarily bad for DBS. After all, this is what DBS needs to do to spread its wings and *plant the national flag* outside Singapore. Hong Kong is the gateway to China, is the *terra firma* for Singapore, Inc. (Quoted in Tan 2001; first emphasis added)

The geography of DBS's business investments was determined not only by its commercial interests, but also by the state's economic strategies.

DBS's Hong Kong expansion not only realized the state's extraterritorial economic priorities, but also reinforced the financialization of the bank by integrating it into increasingly important networks of financial centres and capital markets with the necessary depth to sustain growth across an array of financial activities.

While DBS is a commercial enterprise, its inextricable ties to the state also enable it to produce its own form of autonomy from the 'market', an exceptionalism that magnifies its impact on Singapore's banking reforms and economic growth strategy. DBS disrupted the landscape of finance in Singapore by ushering in new financial logics, but it did so through circuits of influence associated with its links to government rather than through market-driven convergence. The connections between the state and DBS are also continually strengthened by a revolving door of board members and senior positions in the civil service, a financial governing elite (Woo 2014). With DBS, it is difficult to distinguish between corporate decision and state initiative, and DBS's experience through the period of banking reforms demonstrates the co-constituted nature of capitalist institutions that are often irreducible to the market or the state. Instead, we witness the entanglement of both state and commercial interests in the pursuit of capitalist expansion and economic development.

Oversea-Chinese Banking Corporation: Rocky road to reform

OCBC emerged from a coalition of Chinese-family banking firms in 1932 and exemplified the power of *guanxi*-capitalism, a relationship-based capitalism embedded in strong kinship and ethnic ties (Olds and Yeung 1999). Nevertheless, the diffusion of non-Chinese business practices, transnational capital flows and regulatory pressure from institutions like the BIS and the WTO led to significant business restructuring. As MAS introduced new regulatory structures, OCBC, the bank deemed 'Solid as a Rock' (Loh et al. 2000), found itself on shifting sands. Unlike DBS, which made a stratagem of creating a new financial holding company, OCBC found the requirement to divest itself of non-financial assets highly problematic as they were a fundamental component of its business strategy since 1942. Tan Chin Tuan, who was the managing director of OCBC from 1942 to 1972 and chairman from 1966 to 1983, organised the bank with a *keiretsu*-type corporate structure[6] centered on a family-controlled bank (Figure 9.3). Tan saw this as a means of diversification and building 'synergies of scale' (Loh et al. 2000) with other

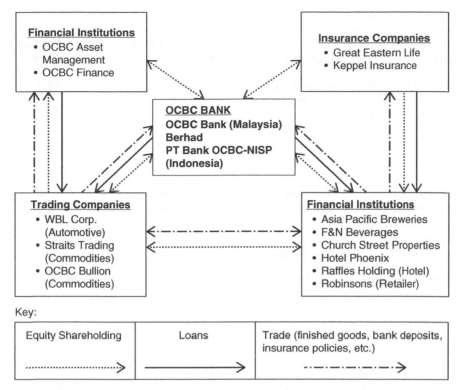

Figure 9.3   OCBC organizational structure until 2004 (mandated deadline for sale of non-core assets). *Sources*: Loh et al. 2000, p. 191; OCBC annual report 2002, pp. 113–19; Low 2002.

businesses, enabling those under the OCBC 'umbrella' to gain from cost efficiencies and access capital at favourable rates.

While the developmental role of state-owned banks such as DBS was more obvious, smaller family banks like OCBC were also vital to Singapore's early industrial development with their focus on local small-to-medium enterprises (SMEs) (Hamilton-Hart 2002). What emerged was an alliance of corporations built upon a tight network of interlocking directorships and shareholdings, with the bank at the centre of the web of firms. Many of these firms eventually became household names in Singapore, including Robinsons department store, Raffles Hotel and F&N Beverages. Through this structure OCBC left an indelible mark on the economic landscape and abetted the aspirations of Singapore's political elite in achieving industrial development goals.

This organizational structure and business strategy proved successful for many decades, and OCBC became one of the largest and most profitable banking groups in Singapore. It also had strong market positions

Table 9.2   Non-core assets to be sold by OCBC

OCBC non-core asset	Industry	Share percentage held by bank	Market value (S$ million)
Robinsons and Company	Department Store/Retailer	16.84	88.5
Straits Trading	Commodity Trading	13.64	64.5
WBL Corp.	Automotive/ Trader	8.85	36.7
F&N	Beverages	8.75	133.5
Raffles Holding	Hospitality	4.56	48.8
Asia Pacific Breweries	Beverages	3.47	46.6
Great Eastern Life*	Insurance	29	1100
Real Estate	Real Estate		1200

* Or Great Eastern Life could sell its 6.2 per cent share in OCBC to avoid cross-shareholding.
Source: Tan 2002.

in Malaysia and Greater China, focused largely on traditional consumer and commercial banking activities. These gave OCBC a large amount of cash on hand with which to buy other firms, including the government-linked Keppel TatLee Bank in 2001. OCBC hoped the acquisition would enable it to reclaim its position as the largest bank before the DBS–POSB merger, but the UOB–OUB merger prevented it. OCBC was thus in the unenviable position of being both the smallest bank in Singapore, and the bank with the most non-core assets to divest by the 2004 deadline (Low I. 2002) (see Table 9.2). OCBC tried to brush off concerns with its CEO arguing that '[its size] certainly does not convey the value and strength of our franchise and our competitive capabilities' (OCBC Annual Report 2001: 9). However, when set against the discourse of size that was pervasive during the period of industrial restructuring, this proved unconvincing; bigger was seen as better for withstanding competitive pressure and for geographical expansion into regional markets. This challenge would continue to fuel rumours over a possible DBS–OCBC merger in the following years. When MAS announced in 2000 that banks would have to relinquish their stakes in non-core business larger than 10 per cent of the firm's assets by the extended deadline of 2002 (later extended again to 2004), it signalled the end of the cross-holding pattern of organizational structures upon which OCBC's business was built. By highlighting the 'maze of bank cross-shareholdings in which not even the bank knew what it owned in relation to what the family owned' (Ong 2002), MAS sent an explicit message to the two major privately owned banks (UOB and OCBC) that their days as 'family banks' were numbered.

Prior to reorganization, OCBC was characterized by two principal profit strategies: 1) principal equity investment in local Singaporean firms (the largest of which were non-financial); and 2) consumer and commercial deposits and loan intermediation, particularly its SME loan business in Singapore, Malaysia and Greater China. The new strategy for OCBC, first spelled out in 2002 as 'New Horizons', was introduced by a new CEO David Conner, who previously led Citibank India and Citibank Japan. Under Conner's leadership from 2002 to 2012, New Horizons targeted three profit streams: 1) wealth and asset management expansion; 2) bancassurance; and 3) tighter integration of Malaysian and Singaporean banking markets (OCBC Annual Report 2002: 10; Connor 2003; 2006). The term used in the 2013 Annual Report – 'comprehensive' – is significant because it justified the packaging of the bread-and-butter retail banking activities with new fee-based products. The emphasis on 'even more entrenchment as a community bank' (OCBC Annual Report 2002: 10) would integrate existing 'community bank' functions with the 'New Horizons' (financialized) strategy, and regain the bank's 'solid as a rock' status (Loh et al. 2000) in its core markets of Singapore and Malaysia, from which new overseas acquisitions could be mobilized. From this strengthened position, the bank then transferred new financial products and practices out through the *guanxi*-network it had built up over eighty years. The OCBC financial holding company now includes one of Southeast Asia's largest asset managers (Lion Global Investors) and one of Southeast Asia's largest insurers (Great Eastern Holdings), both of which have been thoroughly integrated into its banking business, with products channelled through retail banking networks and a private banking arm (Bank of Singapore) (OCBC Annual Report 2013).[7]

Conclusion

While both OCBC and DBS have expanded into private banking (part of a wider industry trend in Singapore), OCBC's strategy has been a retail-oriented financialization, in which traditional retail banking is connected with diversified financial products (e.g., unit trusts, insurance products) tied into wider financial markets. This currently accounts for 45 per cent of revenue growth (OCBC Annual Report 2013). In comparison, DBS has engaged in an institution-oriented financialization, focusing more on commercial banking and institutional clients, with profits accruing from the expansion of investment banking and trading based income.[8] DBS and OCBC have displayed distinctive business responses to banking reforms and government-sanctioned strategies for regional expansion, even as both firms have transformed themselves from banks to financial services corporations. By incorporating finance more fully into the analysis of capitalist change, a variegated financial capitalism approach offers greater topological finesse in revealing how different organizational structures, institutional histories and state–firm relationships may come together, disperse or realign to bring about financialized modes of capitalist growth and their economic outcomes.

The state-sponsored financialization of Singaporean banks has also resulted in broader social impacts with changing consumer financial practices. As banks like DBS and OCBC shift their business focus from loan mediation to fee-paying activities (e.g., investment products and services, bancassurance) in the long shadow of the Asian financial crisis, they connect households and individuals as investors into global financial markets in ever more complex ways. While the financialization of these everyday consumers as 'citizen subjects' contribute towards the expansion of Singaporean banks and the broader developmental objectives of Singapore's (IFC) aspirations (Lai and Tan 2015), it introduced new forms of vulnerabilities that became evident during the 2008 global financial crisis. The bankruptcy of Lehman Brothers in 2008 led to the default of several credit-linked structured notes (collectively known as Minibonds) sold to retail investors in Singapore. The Minibonds (worth almost US$400 million) were issued by ten financial institutions in Singapore, including DBS and the securities arms of OCBC and UOB. The impact of financial losses and accusations of misconduct by distributors prompted public outrage and a level of social activism uncommon in Singapore (Lai 2013b). The Minibonds crisis exposed serious shortcomings in the corporate governance of affected banks and regulatory frameworks regarding the sale of structured products. While subsequent regulatory reforms improve the professional conduct and standards of the financial advisory process and provide some safeguards for vulnerable investors, the result is certainly not a roll back of financialization processes as the emphasis is still on creating 'knowledgeable' and 'responsible' financial subjects who may benefit from wider investment options but who must also shoulder the consequences of their financial decisions (Lai 2016). The reforms following the global financial crisis can be seen as more or less 'technical' fixes to policies whose fundamentals have been left untouched since the Asian financial crisis. The role of finance in advancing the political-economic objectives of the state remains.

This chapter has focused on the role of financialization and its variegated configurations in geographically and historically specific capitalist processes and formations. Scrutinizing the role of the state through its directed financialization of local firms deepens our understanding of capitalist dynamics, financialized modes of production and state–firm relations in what we call variegated financial capitalism. The vital role of banking and finance in Singapore's economic strategy is inextricably tied to multiscalar processes, with the influence of international bodies (such as BIS and WTO), national economic concerns and regionalization strategies of banks. Drawing upon the idea of variegated capitalism, we demonstrate how the transformation of local banks from banking institutions to finance firms reveals the dynamic and multilayered nature of capitalist processes and the interdependence of actors and institutions in developmental goals. Unveiling the entanglement of extraterritoriality and national development goals in this process of state-structured financialization also pushes us beyond the nation-state as an analytical container, or financialization as a unilateral process; instead, we see the multiscalarity of the state and complex state–firm relations as necessary in the production of variegated financial capitalism.

The notion of variegated financial capitalism mobilized in this chapter offers a relational approach to socio-spatial formations, by connecting narratives of 'global' financial integration and the autonomy of (national) financial systems that dominates much of the VOC literature. This emphasizes both the tendential elements of financializing modes of capitalism and its specificity – its limits (cf. Christophers 2015) and extensions – such that its unfolding and its impacts must be taken as conjunctural rather than resultant formations. Variegated financial capitalism does not necessarily provide a new formal category of analysis or phenomenological term, but rather a composite framework that enables us to better situate the spatiality of finance and financial actors – including the state – within the wider frame of capitalism and capitalist variegation. We therefore use the term variegated financial capitalism to open up analytical space for more precise conceptualization of financialization and its role in reshaping the ecologies of capital and power relations within capitalism. Furthermore, as was suggested in the introduction to this chapter, this approach enables us to avoid the fetishization of the 'global' financial crisis as a singularly important moment of change, but instead to historicize such an event in different places as contingent to the variegated processes of financial capitalism.

We have attempted to do this by presenting the dynamic and heterogeneous ways this occurs via the liberalization and reform of Singaporean banks, embedded in state-driven strategies of economic development. Despite what might seem a convergence in the governance and organizational structures of DBS and OCBC due to compliance with regulatory changes, their strikingly different relationships with the state resulted in distinctive pathways to financialization. The relationship between OCBC and the state is more or less an arm's length one as compared to the symbiotic relationship between DBS and the state. Amidst dominant neoliberal tendencies of market liberalization and capitalist expansion, what emerged is a set of variegated capitalist responses, even when corporate structures appear to converge with regulatory requirements. This echoes Clark and Wojcik (2007) on the uneven geographies of organizational change in Germany, whereby different modes of corporate change have been pursued by business managers despite converging pressures of shareholder value as a model of corporate governance. In Singapore, a financialized banking industry has emerged out of specific state–firm relations and the interdependency of actors and institutions. Though faced with the same set of regulatory changes, the state's shareholder status with DBS allows it to shape the landscape of local banking in Singapore in distinct ways compared to more distant state–firm relations in OCBC's case. This produces a complex ecology of state–firm relations and regulatory frameworks, within which local banks have to reorient themselves into the kind of global financial services corporations deemed desirable for regional competitiveness and IFC status. A variegated capitalism approach enables a multiscalar, polymorphic conception of capitalism that connects more convincingly to macro geographies of economic change, as the title of this book portends. Similarly, the incorporation of financial networks, actors and territories into the study of global production and regional development would better enable us to explain global patterns and flows of value creation and capture (Coe et al. 2014). Uncovering the variegated pathways through which financialization is enacted, emplaced and materialised, particularly where it has not *yet* led to financial crisis, represents one means to such engagements.

Notes

1 The use of the term 'Big Bang' reform by Singaporean authorities and policymakers is particularly evocative of the deregulation of financial markets that swept through the City of London in the 1980s.
2 For more nuanced conceptualization of capitalist varieties, see Coates (2005), Crouch (2005), and Streeck and Thelen (2005).
3 The Economic Development Board is a statutory board that coordinates the industrial policy of the Singaporean government, and acts a promotional agent to facilitate foreign direct investment into Singapore.

4   The EDB then concentrated on promoting inward FDI, and later facilitated Singaporean companies' investment abroad
5   http://www.dbs.com/default.page; accessed 13 May 2015.
6   A *keiretsu* refers to a group of companies with interlocking business relationships and cross-shareholdings, usually centred on a bank and trading company. The keiretsu system of corporate governance has been credited with spearheading the post-war growth of Japan and the global success of Japanese firms particularly in the 1970s and 1980s (see Miyashita and Russell 1994).
7   Lion Global Investors is a wholly owned subsidiary of OCBC and is 70 per cent owned by Great Eastern and serves as its asset manager. Great Eastern is 87 per cent owned by OCBC, up from its earlier holding of 29 per cent in 2002 (OCBC Annual Report 2013). Bank of Singapore was formerly ING Asia Private Banking. It was acquired by OCBC from the Dutch ING Group in 2009.
8   Although it also expanded into retail-oriented products especially through its acquisition of POSB, see Lai and Tan (2015).

References

Arndt, H.W., and H. Hill (eds). 1999. *Southeast Asia's Economic Crisis: Origins, Lessons, and the Way Forward*. Singapore: Institute of Southeast Asian Studies.

Arrighi, G. 2003. The social and political economy of global turbulence. *New Left Review* 20:March–April.

Azhar, S., and K. Lim . 2010. Singapore could do with 2 instead of 3 local banks: Lee. *Reuters*, 25 June. Viewed 24 June 2016, http://in.reuters.com/article/2010/06/25/singapore-economy-property-idINSGE65O0GA20100625.

Brown, R.A. 2006. The emergence and development of Singapore as a regional/international financial center. *The Rise of the Corporate Economy in Southeast Asia*. London and New York: Routledge.

Business Times. 1990. When public boon becomes private woe. 5 April, DBS File. National University of Singapore: OCBC Information Resource Unit.

Campbell, J.L. 2004. *Institutional Change and Globalization*. Princeton, NJ: Princeton University Press,

Christophers, B. 2013. *Banking Across Boundaries: Placing Finance in Capitalism*. Chichester: Wiley-Blackwell.

Christophers, B. 2015. The limits to financialization. *Dialogues in Human Geography.* 5 (2):183–200.

Chua, A. 2011. Overseas Union Bank. *Singapore Infopedia*. Singapore: National Library Board. Viewed 25 June 2016, http://eresources.nlb.gov.sg/infopedia/articles/SIP_1787_2011-02-24.html.

Clark, G.L. 1998. Pension fund capitalism: a causal analysis. *Geografiska Annaler: Series B.* 80 (3):139–57.

Clark, G.L., and D. Wojcik. 2007. *The Geography of Finance: Corporate Governance in the Global Marketplace*. Oxford: Oxford University Press.

Clark, G.L., D. Mansfield and A. Tickell. 2002. Global finance and the German model. *Transactions of the Institute of British Geographers* 27 (1):91–110.

Coates, D. (ed.). 2005. *Varieties of Capitalism, Varieties of Approaches.* Basingstoke: Palgrave Macmillan,

Committee on Singapore's Competitiveness. 1998. *Report of the Sub-Committee on Finance and Bankin.*, Singapore: Committee on Singapore's Competitiveness.

Conner, D. 2003. New horizons: Building a platform for growth. Presentation to media and analysts, OCBC Bank. Viewed 25 June 2016, https://www.ocbc.com/assets/pdf/Business-strategy/OCBC_New%20Horizons%20-%20Presentation%20to%20Media%20and%20Analysts.pdf.

Conner, D. 2006. New Horizons II: Embedding OCBC in the region. Presentation to media and analysts, OCBC Bank. Viewed 25 June 2016, https://www.ocbc.com/assets/pdf/Business-strategy/OCBC_New%20Horizons%20II%20-%20Presentation%20to%20Media%20and%20Analysts.pdf.

Cook, M. 2008. Singapore. *Banking Reform in Southeast Asia.* London: Routledge.

Crouch, C. 2005. Models of capitalism. *New Political Economy* 10 (4): 439–56.

Crouch, C. 2011. *The Strange Non-Death of Neoliberalism.* Cambridge: Polity Press.

Daniel, E. 2001. The bank that Olds created. *The Asian Journal* 27.

DBS Annual Report. Various issues. Singapore: DBS Bank. Viewed 25 January 2016, https://www.dbs.com/investor/annual-reports/default.page.

Deeg, R., and G. Jackson. 2007. Towards a more dynamic theory of capitalist variety. *Socio-Economic Review* 5 (1):149–79.

Dixon, A.D. 2011. Variegated capitalism and the geography of finance. *Progress in Human Geography* 35 (2):193–210.

Duménil, G., and D. Lévy. 2004. *Capital Resurgent: Roots of the Neoliberal Revolution.* Cambridge, MA: Harvard University Press.

Dymski, G. 2009. Financial governance in the neo-liberal era. In *Managing Financial Risk: From Global to Local,* eds G.L. Clark, A.D. Dixon and A.H.B. Monk, 48–68. Oxford: Oxford University Press.

Engelen, E. 2003. The logic of funding European pension restructuring and the dangers of financialization. *Environment and Planning A* 35 (8):1357–72.

Engelen, E. 2008. The case for financialization. *Competition and Change* 12 (20):111–19.

Engelen, E., and M. Konings. 2010. Financial capitalism resurgent: comparative institutionalism and the challenges of financialization. In *The Oxford Handbook of Comparative Institutional Analysis,* eds G. Morgan, J. Campbell, C. Crouch, O.K. Pedersen and R Whitley, 601–23. Oxford: Oxford University Press.

Erturk, I., J. Froud, S. Johal, A. Leaver and K. Williams. 2007. The democratization of finance? Promises, outcomes and conditions. *Review of International Political Economy* 14 (4):553–75.

French, S., A. Leyshon and T. Wainwright. 2011. Financing space, spacing financialization. *Progress in Human Geography* 35 (6):798–819.

Froud, J., C. Haslam, J. Sukhdev and K. Williams. 2000. Shareholder value and financialization: consultancy promises, management moves. *Economy and Society* 29 (1):80–110.

Froud, J., S. Sukhdev, A. Leaver and K. Williams. 2006. *Financialization and Strategy: Narratives and Numbers.* London; New York: Routledge.

Hall, P.A., and D.W. Soskice. 2001. *Varieties of Capitalism: The Institutional Foundations of Comparative Advantage.* Oxford: Oxford University Press.

Hall, S. 2013. Geographies of money and finance III. *Progress in Human Geography* 37 (2):285–92.

Hamilton-Hart, N. 2000. The Singapore state revisited. *The Pacific Review* 13 (2):195–216.

Hamilton-Hart, N. 2002. *Asian State, Asian Bankers: Central Banking in Southeast Asia.* Ithaca: Cornell University Press.

Helleiner, E. 1995. Explaining the globalization of financial markets: Bringing states back in. *Review of International Political Economy* 2 (2):315–41.

Henderson, J. 1999. Uneven crises: institutional foundations of East Asian economic turmoil. *Economy and Society* 28:327–68.

Ho, K. 2009. *Liquidated: An Ethnography of Wall Street.* London: Duke University Press.

Howell, C. 2003. Varieties of capitalism: And then there was one? *Comparative Politics* 36 (1):103–24.

Jayasuriya, K., and G. Rodan. 2007. Beyond hybrid regimes: More participation, less contestation in Southeast Asia. *Democratization* 14 (5):773–94.

Krippner, G. 2005. The financialization of the American economy. *Socio-Economic Review* 3 (2):173–208.

Krippner, G. 2012. *Capitalizing on Crisis: The Political Origins of the Rise of Finance.* Cambridge, MA: Harvard University Press,

Lai, K.P.Y. 2013a. Singapore's economic landscapes. In *Changing Landscapes of Singapore: Old Tensions, New Discoveries,* eds E. Ho, C.Y. Woon and K. Ramdas, 196–217. Singapore: NUS Press.

Lai, K.P.Y. 2013b. The Lehman Minibonds crisis and financialization of investor subjects in Singapore *Area* 45 (3):273–82.

Lai, K.P.Y. 2016.Financial advisors, financial ecologies and the variegated financialization of everyday investors. *Transactions of the Institute of British Geographers* 41 (1):27–40.

Lai, K.P.Y. 2017. Financialization of everyday life. In *The New Oxford Handbook of Economic Geography,* eds G. Clark, M. Feldmann, M. Gertler and D. Wojcik. Oxford: Oxford University Press.

Lai, K.P.Y., and C.H. Tan.2015. Neighbors first, bankers second: Mobilizing financial citizenship in Singapore. *Geoforum* 64:65–77.

Langley, P. 2008. *The Everyday Life of Global Finance: Saving and Borrowing in Anglo-America.* Oxford: Oxford University Press.

Lee, H.L. 1998. Fund management in Singapore: New directions. Speech by Deputy Prime Minister Lee Hsien Loong at The Investment Management Association of Singapore Seminar, 28 February. Viewed 25 June 2016, http://www.mas.gov.sg/news-and-publications/speeches-and-monetary-policy-statements/speeches/1998/fund-management-in-singapore-new-directions--26-feb-1998.aspx.

Lee, H.L. 2000. Measures to separate financial and non-financial activities of banking groups. Speech by DPM Lee Hsien Loong at the Association of Banks in Singapore Dinner, 21 June. Viewed 25 June 2016, http://www.mas.gov.sg/news-and-publications/speeches-and-monetary-policy-statements/speeches/2000/measures-to-separate-financial-and-non-financial-activities-of-banking-groups--21-june-2000.aspx.

Lewis N., W. Larner and R. Le Heron. 2008. The New Zealand designer fashion industry: making industries and constituting political projects. *Transactions of the Institute of British Geographers* 33 (1):42–59.

Lim, C.Y. 2011. The National Wages Council (NWC) and macroeconomic management in Singapore. In *Crisis Management in Public Policy: Singapore's Approach to Economic Resilience*, eds W.M. Chia and H.Y. Sng, 3–18. Singapore: World Scientific Publishing.

Liow, E.D. 2012. The neoliberal-developmental state: Singapore as case study. *Critical Sociology*. 38 (2):241–64.

Loh, G., C.B. Goh and T.L. Tan. 2000. *Building Bridges, Carving Niches: An Enduring Legacy*. Oxford: Oxford University Press.

Low, I. 2001. DBS says sorry to UOB and OUB. *The Straits Times*, 1 August, DBS File, OCBC Information Resource Unit, National University of Singapore.

Low, I. 2002. It has always been book value, says MAS. *The Straits Times*, 7 August, DBS File, OCBC Information Resource Unit, National University of Singapore.

Low, L. 2001. The Singapore developmental state in the new economy and polity. *The Pacific Review* 14 (3):411–41.

Martin, R. 2002. *Financialization of Daily Life*. Philadelphia: Temple University Press.

Miyashita, K., and D. Russell.1994. *Keiretsu: Inside the Hidden Japanese Conglomerates*. New York: McGraw-Hill.

Monetary Authority of Singapore. 2012. *Sustaining Stability: Serving Singapore*. Singapore: Monetary Authority of Singapore.

Ngiam, T.D. 2011. *Dynamics of the Singapore Success Story: Insights by Ngiam Tong Dow*. Singapore: Cengage Learning Asia.

OCBC Annual Report. Various issues. Singapore: OCBC Bank. Viewed 18 January 2016, https://www.ocbc.com/group/investors/annual-reports.html.

Olds, K., and H.W.C. Yeung. 1999. (Re)shaping 'Chinese' business networks in a globalizing era. *Environment and Planning D: Society and Space* 17 (5):535–55.

Ong, C. 2002. Untangling the maze of bank cross-shareholdings. *Business Times*, 9 August, OCBC File, OCBC Information Resource Unit, National University of Singapore.

Ong, C.T. 2004. Basel and Beyond. Opening Address at the 2004 ISDA Regional Conference, 26 October. Viewed 25 June 2016, http://www.mas.gov.sg/news-

and-publications/speeches-and-monetary-policy-statements/speeches/
2004/2004-isda-regional-conf-basel-and-beyond.aspx

Peck, J. 2005. Economic sociologies in space. *Economic Geography* 81 (2):
129–75.

Peck, J. 2013. Explaining (with) neoliberalism. *Territory, Politics, Governance*
1 (2):132–57.

Peck, J., and N. Theodore. 2007. Variegated capitalism. *Progress in Human
Geography* 31 (6):731–72.

Peck, J., and N. Theodore. 2010. Mobilizing policy: Models, methods, and
mutations. *Geoforum* 41 (2):169–74.

Perry, M., L. Kong and B.S.A Yeoh. 1997. *Singapore: A Developmental City
State*. New York: Wiley.

Phelps, N.A., and F. Wu. 2009. Capital's search for order: Foreign direct
investment in Singapore's overseas parks in Southeast and East Asia. *Political
Geography* 28 (1):44–54.

Pike, A., and J. Pollard. 2010. Economic geographies of financialization.
Economic Geography 86 (1):29–51.

Poon, J.P.H., and E.R Thompson. 2001. Effects of the Asian financial crisis on
transnational capital. *Geoforum* 32:121–31.

Prasad, M. 2012. *Land of Too Much: American Abundance and the Paradox of
Poverty*. Cambridge, MA: Harvard University Press.

Purushothaman, R. 1998. Seems like DBS got POSBank on silver platter. *The
Straits Times*, 21 August, DBS File. OCBC Information Resource Unit,
National University of Singapore.

Raj, C. 1999. Market hails DBS' proposed revamp. *Business Times*, 29 June,
DBS File. OCBC Information Resource Unit, National University of Singapore.

Rodan, G. 1989. *The Political Economy of Singapore's Industrialization*.
London: Macmillan.

Rodan, G. 2012. Consultative authoritarianism and regime change analysis. In
Routledge Handbook on Southeast Asian Politics, ed. R. Robinson, 120–134.
London: Routledge.

Stiglitz, J. 2003. Globalization and the economic role of the state in the new
millennium. *Industrial and Corporate Change* 12 (1):3–26.

Streeck, W., and K. Thelen (eds). 2005. *Beyond Continuity: Institutional Change
in Advanced Political Economies*. Oxford: Oxford University Press.

Stubbs, R. 2009. What ever happened to the East Asian Developmental State?
The unfolding debate. *The Pacific Review* 22 (1):1–22.

Sub-Committee on Banking and Financial Services. 1985. *Report of the Sub-
Committee on Banking and Financial Services*. Singapore: The Sub-Committee.
on Banking and Financial Services

Sunley, P. 2008. Relational economic geography: A partial understanding or a
new paradigm? *Economic Geography* 84 (1):1–26.

Tan A. 2001. S'pore Inc. wins, but not minorities. *Business Times*, 13 April.

Tan, C.H. 2005. *Financial Markets & Institutions in Singapore*, 11th edn.
Singapore: Singapore University Press.

Tan, K. 2002. Clock ticking for S'pore banks. *The Edge Singapore*, 19 March, OCBC File. OCBC Information Unit, National University of Singapore.

Tan, C.H., and J.Y.S. Lim. 2007. *Singapore and Hong Kong as Competing Financial Centers*. Singapore: Saw Centre for Financial Studies.

Van der Zwan, N. 2014. Making sense of financialization. *Socio-Economic Review* 12 (1):99–129.

Weiss, L. (ed.). 2003. *States in the Global Economy: Bringing Domestic Institutions Back In*. Cambridge: Cambridge University Press.

Williams, K. 2000. From shareholder value to present-day capitalism. *Economy and Society* 29:1–12.

Wójcik, D. 2007. Geography and the future of stock exchanges: Between real and virtual space. *Growth and Change* 38 (2):200–23.

Woo, J.J. 2014. *Converging Interests, Emerging Diversity: A 'Nested Instrumental Approach' to Financial Policy in Hong Kong, Singapore and Shanghai*. Unpublished dissertation, Lee Kuan Yew School of Public Policy, Singapore.

Yeung, H.W.C. 2000. State intervention and neoliberalism in the globalizing world economy. *The Pacific Review* 13 (1):133–62.

Yeung, H.W.C. 2004. *Chinese Capitalism in a Global Era: Towards Hybrid Capitalism*. London: Routledge

Yeung, H.W.C. 2012. East Asian capitalisms and economic geographies. In *The Wiley-Blackwell Companion to Economic Geography*, eds T.J. Barnes, J. Peck and E. Sheppard. Chichester: John Wiley & Sons.

Index

Page references to Figures or Tables will be followed by the letters 'f' or 't' in italics as appropriate, while references to Notes will be followed by the letter 'n' and Note number.

Money and Finance After the Crisis: Critical Thinking for Uncertain Times,
First Edition. Edited by Brett Christophers, Andrew Leyshon and Geoff Mann.
© 2017 John Wiley & Sons Ltd. Published 2017 by John Wiley & Sons Ltd.

Printed and bound by CPI Group (UK) Ltd, Croydon, CR0 4YY

27/10/2024

14580160-0005